Management of
System Engineering

Management of System Engineering

Wilton P. Chase
TRW Systems

Robert E. Krieger Publishing Company, Inc.

Malabar, Florida

Original Edition 1974
Reprint Edition 1984

Printed and Published by
ROBERT E. KRIEGER PUBLISHING COMPANY, INC.
KRIEGER DRIVE
MALABAR, FLORIDA 32950

Printed in the United States of America

Library of Congress Cataloging in Publication Data

Chase, Wilton P. (Wilton Perry), 1908 -
    Management of system engineering.

    Reprint. Originally published: New York: Wiley, 1974.
(Wiley series on systems engineering and analysis)
    Bibliography: p.
    Includes index.
    1. Systems engineering.   I. Title.
TA168.C47   1984   620.7'068   83-18746
ISBN 0-89874-682-5

SYSTEMS ENGINEERING AND ANALYSIS SERIES

In a society which is producing more people, more materials, more things, and more information than ever before, systems engineering is indispensable in meeting the challenge of complexity. This series of books is an attempt to bring together in a complementary as well as unified fashion the many specialties of the subject, such as modeling and simulation, computing, control, probability and statistics, optimization, reliability, and economics, and to emphasize the interrelationship between them.

The aim is to make the series as comprehensive as possible without dwelling on the myriad details of each specialty and at the same time to provide a broad basic framework on which to build these details. The design of the books will be fundamental in nature to meet the needs of students and engineers and to insure they remain of lasting interest and importance.

Preface

There have been several excellent treatises published on systems engineering. They cover both the process and the procedures for design and development of complex equipment systems. Collectively, they provide a good description of systems engineering technology. However, very little has been written on the dynamic process by which a system design is created.

It is very much easier to describe the methods and techniques which can be employed for producing particular system end products than it is to explain the creative and decision-making processes which underlie the selection of given technologies and design disciplines for application in a given system design. It involves the age-old distinction of possessing insight and understanding of a problem and of how to approach its solution as against acquiring a rote knowledge of how to apply given techniques to implement a solution once it is defined.

A background in experimental psychology helps provide a basis for understanding the creative and decision-making process which occurs in conceiving, designing, and developing a system. But it is unusual for a behavioral scientist with such a background to penetrate the equipment-oriented world of system engineering and to work intimately with system project personnel in making the day-by-day decisions which progressively accumulate to compose a system design. The writer considers himself fortunate that he has had such an opportunity, not just once, but several times, involving systems of varying degrees of complexity. In the process, he has been able to maintain a professional interest in studying the systems design and engineering process. Because of this interest, he has been able to contribute to the development and application of system engineering methodology in relation to complex large-scale military and space systems. Throughout it all, he has tried to keep an objective viewpoint, even though at times he has been deeply involved emotionally in trying to sell improvements in methodological approaches for given system design and development projects.

The background for this book, therefore, is the accumulation of observations and experiences over a considerable period of time. It attempts to reveal that arriving at a system design is not as straightforward and logical a process as the published descriptions of system engineering methods and techniques could lead one to believe. In describing how organized groups of presumably technically qualified people create designs, we will, of course, be describing various scientific methods and techniques available to them for enhancing the capability for making sound system design decisions. Decisions there will be, but whether the best available tools will be employed to aid in making them is a matter of choice. Describing the elements and the climates which influence how system design decisions are made is what this book is all about. It is not necessarily a direct path from a problem to a given solution.

There is a word of caution for the prospective reader. If he thinks that he understands how the basic system design and engineering process takes place and is looking for help in applications of specific techniques, or for a "cookbook" procedure describing the steps for "how to do it," then this is not the book for him. It is suggested that he consult the bibliography at the conclusion of the text to see if there are any books dealing with systems engineering with which he may not be familiar.

It is further emphasized that this book has been written with the intention only to explain and describe the basic system design and engineering process and to help the reader develop a realistic viewpoint in regard to it. There is nothing in it that can be rigidly interpreted or directly applied to a specific system design and development project. Each such project presents unique requirements for selection of methodological and technological approaches. We will attempt only to describe and explain basic concepts and principles which can be employed as a foundation upon which effective system design and engineering approaches can be built.

Los Angeles, California *Wilton P. Chase*

Contents

1

Acquiring the System Viewpoint

What Is a "System"?

Every time you breathe, lift a fork with food to your mouth, drive your automobile, cash a check, or do any of the multitude of things which occur in everyday life, you are functioning as an integral part of some system. The given system in which you are functioning at the moment depends upon your particular perspective. In fact, most of the time you would have to consciously force yourself to think of the role you are playing as a part of a system. In other words, what constitutes a system is a state of mind. Consequently, the concept of a "system" is an abstract, devised, synthetic entity. This is true whether it be one which has been invented and manufactured by man or one which exists in nature as a self-contained physical or biological entity. In either case, it is a "system" only because someone views it from a given point of reference. He sees an organization or an integration of forces or events for which he can define a set of boundaries. Also, he can explain to his own satisfaction what the energy transformations are which must occur in order to account for, or to attain, a predictable outcome under controlled conditions. As such, a system can only be described in some form of language, such as pictographs, words, diagrams, motion pictures, mathematical and statistical formulas, or—in the most modern language form of all—computer programs. Hopefully, a system can be precisely described and its description accurately communicated to whoever must deal with it. However, it is virtually impossible for any two individuals to achieve a common understanding of a given system.

From the individual's standpoint, he can either be subjugated within a system, and function only as a performing element of it, or he can control and use a system to achieve his desired objectives. It is the development of systems to provide this latter capacity that concerns us here. Further, we shall be concerned only with the design of man–machine systems and not the man–man or man–paper systems which constitute our political, social, and economic milieu. However, we will have to be concerned with such systems, to the extent that they furnish the climate which either stimulates or constrains the creation of effective man–machine systems. We will also be concerned tangentially with natural systems, both physical and biological, because it is the scientific understanding and the mastery of the use of such systems to achieve desired outcomes that constitute the basic technology upon which effective man–machine systems can be constructed. This is true whether we should choose to consider as an example the earliest domestic system of primitive man employing stone implements, medieval weapon systems, transportation systems based on the steam engine or the airplane, or the most complex spaceship for manned exploration of distant planets.

An example of complex problem solving in an early culture which involved using a "system approach" by a people who had neither a written language nor the technology of the wheel is described by Victor W. von Hagen in "Realm of the Incas"* as follows:

> Agriculture was bound up very closely with terracing and irrigation, since the amount of flat land was severely limited and the Andean valleys are deep and narrow. The sides of the valleys were wonderfully terraced, and it is these that excite when seen for the first time. The rainy season run-off carries away soil; terracing prevented it. Terracing further extended the soil community where the exiguous earth surfaces lacked for want of space; so the Indians, too, were soil-makers.
>
> Under Inca rule, terracing of the Andean valleys was a systematic part of their methods of soil preservation and soil creation. In the

*V. W. von Hagen, *Realm of the Incas*, New American Library of World Literature, Inc., New York, 1957, pp. 70–72.

greater projects, those, for example, of Pisac—where the terraces stand poised over the heights of the Vilcanota River, or, for another example, at Ollantaytambo (where the workers cut into the living rock)—professional architects were sent out from Cuzco to plan them. [We can substitute the contemporary appellation, "system engineer" for "architect" in the dictionary sense of "one who plans and achieves a difficult objective."] Victor von Hagen continues: It was an enormous expenditure of labor. That these terraces still stand after five centuries is sufficient testimony to the foresight of those Inca rulers.

Irrigation was tied closely with terracing and so naturally with agriculture. It was the life-blood of empire. In the wet season rain does not always fall nor does all this borax-filled earth hold the water, so irrigation was the answer, and the Inca engineers harnessed the brawling streams pouring out of glaciers and brought them down in the most careful manner to water fields, even though separated by immense distances from the water-course. These techniques helped to control the density of population and gave the social body a meticulous balance between population and productivity. Much Inca-directed skill was devoted to irrigation. There were immense water reservoirs in the fortress of Sacsahuaman above Cuzco; water was laid underground in superbly made stone-laid sewers in many widely spaced areas. Rivers were straightened, canalized as one sees the Vilcanota River, a few miles east of Cuzco and below the great fortress of Pisac. This type of advanced engineering once extended throughout the empire, but is now only dimly seen since so much has been lost to the insults of time.

Irrigation techniques are inseparable from a developed agriculture, and their elaboration marked man as a settler, brought about a corporate life with settled habits; it also created the city-state.

Irrigation, it need hardly be said, was not an Inca invention but it *was* an Inca perfection. On the desert coast the Mochicas, to mention one great pre-Inca dynasty, had vast irrigation works, and their heirs, the Chimús, built their cities on the ruins of those of the Mochicas and extended the irrigation system so that their cities were supplied by gigantic stone-laid reservoirs. And far to the south of Peru in the Ica-Nasca regions, these cultures, between the years 500–800 A.D., also elaborated immense irrigation works; the underground reservoirs are still known as *puquios*. Of all this, the Incas were the inheritors.

Yet under the Incas terracing was perfected and extended. Water was so engineered as to be induced at the top of the terraces, thence it ran down from one gigantic terraced step to another, the whole being watered by a single stream. Now water conduction demands careful design and must be determined by a knowledge of hydrographic conditions, the nature of the soil, and the general conformation of the land. To secure the flow it must run down a slight incline: too fast, it will erode the banks; too slow, it will allow weeds to grow and silt will choke up the channels. It is scarcely curious that wherever in this variegated globe water conduction was practiced, the techniques of it are almost identical. In ancient Mesopotamia, after its conquest in 1760 B.C. by Hammurabi, land exploitation was centralized, which led naturally to the erection of canals, reservoirs, and irrigation dykes. King Hammurabi's "Water Code" is written in a form that sounds like an Inca text: "Each man must keep his part of the dyke in repair." Royal letters were dispatched to governors giving them over-all responsibility to keep the waterways open and the dykes in repair: "Summon the people who hold the fields on the side of the Damanu Canal that they may scour it."

This is only one instance of parallelism in human inventions. For after all it is the stomach that is the all-compelling motive of invention, and man through his ages tested, tasted, and tried the fruits and grains that fell within his ken; he learned to evaluate them, select and plant them—for this reason one need not become unduly overexcited about the parallelism of means and methods in terracing used by peoples out of contact, for where geographical conditions are similar, agricultural methods and water induction will also spring up naturally. One need not fall back on diffusion as an explanation. The peasant is land-bound. He may be a Peruvian *puric* or an Egyptian *fellah*, but he remains the "eternal man." He is, in the Spenglerian dictum: "the eternal man, independent of every culture that ensconces itself in the cities. He precedes it, outlives it: a dumb creature propagating himself from generation to generation, soil bound, a mystical soul, a dry, shrewd understanding that sticks to practical aptitudes."

The Scientific Approach to System Design and Engineering

Man creates systems to achieve some extension of his capability for sensing and perceiving physical phenomena; or to process and assimiliate multifaceted information for thinking and decision making; or to increase his physical prowess for going places and doing things ever more quickly, precisely and powerfully.

Table 1 has been compiled to show that the basic steps in the problem-solving process of designing and engineering ways of enhancing or extending our basic capabilities have not changed since man first evolved the ability to shape or make things in order to solve basic problems of satisfying his physiological needs. Only our knowledge of the nature of the available resources and how to utilize them has increased. Complex large-scale systems such as we are designing and developing in the electronics and space age are the result of the increase in this knowledge which enables us to harness and integrate the available energy and the inherent capabilities of various combinations of physical, mechanical, electrical, thermal, chemical, physiological elements so that they function as a coherent unitary operational process to achieve a given purpose.

To use an exact technical term, systems are *anthropocentric.* In fact, they tend to reflect human qualities in their designed characteristics, sometimes to the extent that their performance capability suffers because more appropriate "nonhuman" ways of accomplishing something were not conceived. In other words, even though all systems are anthropocentric in their conception and utilization, designing them to possess anthropomorphic characteristics may not necessarily result in an optimum design for the work to be accomplished.

The basic steps involved in designing and proving a system, whether they be followed consciously or unconsciously, are shown in the extreme right-hand column of Table 1. The systematic approach to apply fully system engineering methodology and techniques to derive a coherent system design is depicted in Figure 1. Admittedly, this is a very "busy" chart, but it presents for study and discussion all the types of parameters which must be considered. Specific parameters must be determined as applicable to a given system design problem. Arriving at a system design and then accomplishing the engineering required to translate the design into equipment, facilities, procedures, trained personnel, and logistics support must always be accomplished in relation to specific requirements. System design and engineering cannot be accomplished as a generalized exercise in a vacuum.

The Problem of Communicating About the System Approach

There are tremendous language difficulties to be overcome in effectively communicating system concepts and in describing the systems approach. Mathematical modeling has been attempted as a means of substituting quantitative descriptions for qualitative ones. However, verbal language forms are more basically ingrained in human behavior for carrying out thought processes and for making decisions than are mathematical or statistical languages. The latter must be translated into verbal terms for effective employment in decision making. Consequently, in the process of arriving at system designs, mathematical modeling exercises are only advisory at best, and more often than not are "gold plating" of what is already obvious to the decision maker.

Properly understood, mathematical modeling is a useful tool for studying performance characteristics that will result from different quantitative input–transformation–output performance combinations (provided they can be achieved functionally in man–equipment designs), especially when they can be studied through simulation in computer programs. However, the knowledgeable and imaginative decision maker is required to employ the output of such simulations only as an aid in deriving a given system design. He has a choice in particular applications of ignoring them completely, using them appropriately and wisely, or overusing them in such a way that equally significant factors are ignored or go undiscovered because curiosity is stifled. The worst enemy of an effective system design and system engineering process is the uncritical application of modeling programs and

Figure 1. (see foldout opposite) A schematic block diagram illustrating the basic system design and engineering parameters and their interrelationships.

"cookbook" system design and engineering techniques and procedures in lieu of personal understanding and analysis of a design or engineering problem.

As a start in communicating verbally about the system approach, here are some definitions of basic terms:

- *System analysis* is the overall process for arriving at the best mix of equipment, personnel, and procedural requirements for a system design.
- *System definition* is the determination of the integrated quantitative and qualitative requirements for prime mission equipment, its supporting equipment and facilities, procedures (including computer programs), personnel selection and training, and logistics support.
- *System engineering* is the process of selecting and synthesizing the application of the appropriate scientific and technological knowledge in order to translate system requirements into a *system design,* and, subsequently, to produce the composite of equipment, skills, and techniques and to demonstrate that they can be effectively employed as a coherent whole to achieve some stated goal or purpose.

The Gross Steps in the System Design and Engineering Process

Structuring of the thought process, emphasizing the scope of a system problem, and fathoming the details which require consideration in arriving at possible design solutions are minimum basic requirements upon which to build an effective system design and engineering process. Figure 1 depicts the irreducible gross functional steps which must be followed.

In relation to Figure 1, the process of system design always starts with some need. This need is expressed in terms of *values.* They concern the utility of the product, how much it is desired to pay for a given performance, and how quickly it is desired that it be available for use. The expression of these values, whether precise or imprecise, constitutes the *standards* by which the customer evaluates the acceptability of a system design. They are the basis for the expression of his satisfaction or dis-

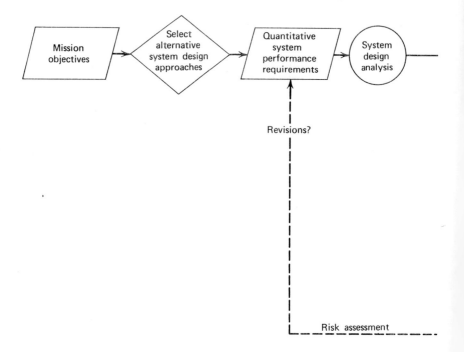

Figure 2.   The basic system design and engineering process.

satisfaction with the product. Without a set of values, a customer has no felt need or set of requirements for a product and no amount of system engineering expertise to improve a system design will cause the product to be marketable.

Given a set of requirements to fulfill in order to satisfy a need for a system product, the next step is to accomplish system analyses. Collectively, these are concerned with deriving a design

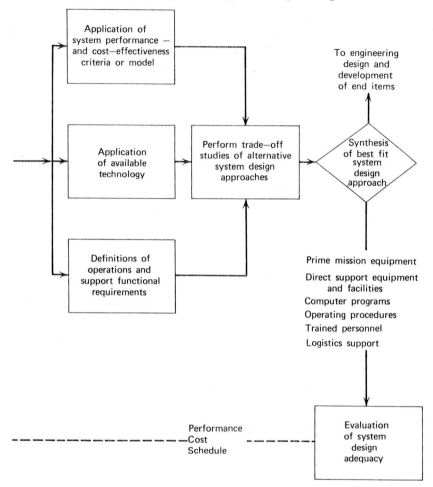

Figure 2. *Continued*

solution to meet the customer's operational requirements. This is true whether he has stated them in whole or in part, or whether it is necessary to develop assumptions about them. In order to arrive at alternative approaches to system design, three lines of analysis and study effort must be pursued. These are shown in the information flow diagram of the basic system engineering process as depicted in Figure 2.

As shown in both Figures 1 and 2, the mainstream effort is always concerned with exploring the applicability of the available science and technology. In most system design efforts, this study will deal with selecting off-the-shelf techniques which can be either directly used or easily modified. Of course, there will be cases in which achieving a desired performance will require inventions. In such an event, it is necessary to make a judgment as to whether the inventions can be achieved within the customer's stated performance, cost, and schedule constraints.

Concurrently, the customer's value standards must be expanded and made more explicit by system performance-effectiveness criteria or, preferably, by developing a qualitative and quantitative model which can be employed in trade-off studies. The system performance-effectiveness model is an integration of figures of merit assigned to describe equipment performance capability and its availability to perform when needed. Availability in turn is an integration of equipment reliability and maintainability. In some systems, such as military ones, survivability/vulnerability is also an important factor in availability. (See analysis breakdown structure under the heading, "Technical Performance Measurement Objectives," in Figure 10.) Such a model cannot be developed completely or independently of the definition of the functional requirements for producing and deploying the system.

Definition of functional requirements is shown as the third line of analytical effort needed to accomplish system design trade-off studies. Detailing the functional sequences in order to describe the operations, control, and maintenance requirements permits systematic derivation of performance requirements for equipment, facilities, procedures, personnel, and logistics support end items. The functional requirements, when used in conjunction with the system performance-effectiveness criteria, furnish the system-oriented performance standards baseline for evaluating alternative system design approaches in trade-off studies. The results of trade-off studies at all levels of detail (for example, selection of subsystems to form a coherent integrated system, selection of components to compose subsystems, and, finally, selection of parts, materials, and processes to fabricate components) furnish the data by which the system management, whether it be exercised by one individual or collectively by a group consensus, can select and synthesize a "best fit" system design. This is the ultimate act of judgment by "choice" which constitutes the creation or invention of a new system product. No procedure can be devised to mechanize the process in order to ensure infallibility. Even modern computerized information processing can only furnish aid in decision making; it cannot substitute for it.

The process of analyzing requirements and of conducting trade-off studies as a basis for selecting design approaches becomes progressively more refined and detailed as the process is iterated. Finally, a *total* and detailed system design solution is reached. The adequacy of the design solution is always a matter for the customer to judge in terms of attaining his objectives in relation to what he considers he can afford and when he can have it to use. From an *operations analysis* viewpoint, a given system design is evaluated from the standpoint of its *cost effectiveness,* which is an integration of the factors of its performance, cost, and utility from the standpoint of the customer's value standards.

Just as arriving at a system design is an iterative process, so is arriving at an understanding of its underlying problem-solving and decision-making process also an iterative process. The remainder of the book will be devoted to further detailing the process as viewed from different angles, with the hope that the reader's understanding will grow and develop to the point that he will echo the great Gestalt psychologist Wolfgang Kohler's cry of "Aha!"—symbolic of insight attained.

2

The Basic System
Design Process

The Human and Organizational Context of System Design

Because of the complexity and scope of modern large-scale systems, a total system design can be accomplished only through the organization of a multitude of human endeavors. Consequently, such a system design and development program, involving large numbers of personnel and a multiplicity of functions, tends to become bureaucratic. A bureaucracy has four main characteristics—a hierarchy of authority, impersonality, specialization, and a set of rules.

The hierarchy of authority is made necessary by the need to achieve efficiency and economy in system management. The numerous scientific, engineering, and business administration personnel responsible for managing the design and engineering of a system are usually the product of training in highly specialized fields. A system project manager with decision-making authority for a complex system design and development program is normally chosen because of his closely allied specialty and technical competence for handling problems which may arise in developing the prime mission equipment. Other factors upon which the successful use of the prime equipment is dependent—namely, support equipment, personnel, facilities, operational procedures, and logistics support—are of less concern to him.

To unify all the separate and diverse scientific, technologican, and administrative efforts required for achieving an effective system design, the successful system engineer or system project manager must be able to convert basic scientific, technical, and management data into a significant understanding of their separate and relevant importance for the total system design. He cannot rely solely on his own background and experience, but must use the talents and advice of competent subordinates. The impersonality inherent in the large organization required to create complex systems makes constant watchfulness necessary in order to ensure that the viewpoints of individuals with the depth of understanding required to make basic system design decisions are not submerged, because they may occupy an inferior status in the hierarchy of decision-making authority.

Watchfulness is also important to ensure that individuals with creative talent for deriving and integrating complex system designs are not lost in the bureaucratic maze, or confined too closely to an area of intensive specialization. Such talent must be located, nurtured, and given an opportunity to function, for here is the source of the breakthroughs that produce the kind of esoteric systems which are in demand.

Finally, the diversities of technologies and the intense specialization required in large system design and development projects make a "set of rules," or formalized procedures for operating in complex organizations, a necessity. These rules or procedures become the basis of communication among the various groups. Only if clear communications among the varied specialized efforts is established can there be an integrated coherent program effort, such as is required to design and develop a system composed of complex subsystems which must function effectively together as a unified entity.

How System Designs Happen in the Real World

The first and foremost basic reality about systems is that they are supposed to be designed to perform an organized set of operations in order to satisfy a defined user requirement. This basic purpose tends to be forgotten because of the bureaucratic division of a system design and development effort and the resultant group concentration upon the many separate supporting technologies. Each technology tends to become functionally an independent organization, even within the context of a presumably integrated program. Unfortunately, such specialization can lead to over- or under-design in particular subsystems at the expense of considerations of the relevance of the subsystem design characteristics for satisfying a given system purpose.

Simple as the steps of the system engineering process appear to be, their implementation is not simple. Not only is it difficult to develop the requisite viewpoint, understanding, and perspective, but, additionally, it requires the courage to stand alone in order to live by a scientific approach when it must be applied outside the confines of a well-defined and politically insulated laboratory environment. In attempting to apply strictly scientific thinking to the design of complex systems, we must contend with the political, social, and economic influences of the customer. The customer's requirement for a system is stated in value terms which reflect such influences. He wants a certain performance outcome by a specified date and within a certain cost estimate.

As a basic fact, it must be emphasized that the various individuals ultimately responsible for a system design possess diverse educational backgrounds and specialized technical experience. Because of these differences in educational backgrounds and job experience, as well as in basic abilities and sensitivity to political, social, and economic pressures, the approach to system design varies widely. Too often problems in technical areas which need resolution in relation to overall system design are not considered objectively, but are considered in relation to a personal fund of available specific solutions, or to the personal biases of key project management personnel, or to the customer. In this circumstance, a design can occur before any attempt to objectively understand and define the system requirements has been made. The result is aptly described as a "random, piecemeal, accidental discovery of the system." The various pieces are accumulated by relatively independent design actions. Then, how to properly integrate the pieces into a smoothly operating system entity becomes the major system engineering task. As the resultant true operational capability becomes apparent, the customer usually expresses grave concern because his value system has been violated. He realizes that he is not going to get what he wants without paying more and waiting longer for it.

Invention of System Engineering Crutches and Their Mythical Utility

When system engineers and system project managers make decisions which lead to succesful design in terms of results obtained by its use, the product is usually characterized by such terms as "imaginative," "useful," "dependable," "economical," "ingenious," "easy to operate and maintain," "clever," and by other adjectives expressing value judgments which connote user satisfaction and admiration. However, when system design decisions result in "glitches" or inadequacies in product performance, there is usually great concern with principles of "system management," "system requirement analysis," "program management," "configuration management," "system engineering management procedures," and other such schemes which purport to organize, document, and provide system engineering management visibility and control of the *total* design process.

With the advent of these highly publicized and formally directed schemes, we encounter some commonly held myths about their utility. Among these are:

1. "If you prescribe a procedure for decision making which incorporates all of the essential logical and scientific steps, you can impose it by management edict as a 'forcing function' in order to obtain a successful system design." Experience has shown that nothing could be farther from the truth. A design cannot be regimented into being a success by enforcing compliance with system engineering management schemes or procedures. There is no doubt that people can be trained to apply these procedures "after-the-fact" in order to describe how design decisions were made. However, unless a designer employs the procedures as his personal process for identifying, organizing, and understanding the interaction of all elements basic to making his decisions, the use of the prescribed procedures is of little value. The "forcing function" which counts is the designer's ability to analyze all the factors, evaluate their relative importance, and select a combination which will achieve a desired objective. Such ability is acquired only through education in the fundamental methodologies and experience in their application

for solving unique problems. In other words, many can imitate the steps in the scientific method, while a few develop an understanding of the basic ideas and a capability for their effective application.

2. "By emphasizing qualitative characteristics of system performance, and by assigning them quantitative values as design requirements (such as operability, reliability, maintainability, safety, value engineering, and so on), you can motivate designers to do a better job." This is true to the extent that you can call attention to the desirability of attaining certain product characteristics from the user's standpoint which must not be ignored, but unless the individual designer possesses an understanding of a given problem, creative ability, and appropriate technical knowledge, no amount of "number assigning" will directly influence the quality of the resultant product.

3. "Employment of a highly structured data reporting scheme will result in more adequate design solutions." Such data reporting schemes are used to progressively describe all aspects of a system from its overall functional performance right down to the use of detailed parts, materials, and processes in an all-encompassing system "performance specification tree." In actuality, poor system designs as well as good designs can be described in infinite detail. The test of effective system engineering is not in the descriptive words and diagrams, but in the achievement of the desired mission objectives by appropriate product performance.

4. A myth of long standing is that if you have enough money and engineers you can solve any design problem within a restricted time period. Of course, from a strictly actuarial or probabilistic standpoint, this approach can lead to satisfactory outcomes because it can result in situations where the right persons will be available when needed. However, there are systems which have been amply funded, and upon which thousands of scientists and engineers have been employed, but have not been brought to successful fruition, either in terms of attaining the originally specified operational performance characteristics, or of meeting the given delivery schedules and cost estimates.

The many attempts to improve the quality of complex system designs by devising "paperwork," "mechanical," or "numerical" approaches to superimpose a *total* system engineering viewpoint upon the basic hardware design process have tended to create an unnecessary and regrettable dichotomy between "creative hardware designers" and "system-oriented engineers." Much effort needs to be applied by both system-oriented engineers and hardware designers to mutually streamline their intergroup communication techniques (or the "set of rules" in the bureaucratic sense) in order to serve their common objectives without benefit of any middleman group of "system engineering specialists," and without a need for crutches.

The Doctrine of Fallibility

No two responsible system designers, operating in the same bureaucratic environment, supplied with the same set of criterion values and the same technological applications trade-off information to evaluate, and presumably acting reasonably and logically, will arrive at identical conclusions or formulate identical concepts concerning what courses of action should be taken.

Theoretically, as we have been emphasizing, since synthesizing an effective system design is essentially an objective application of the scientific method, then with all the facts in hand, the final decisions establishing the parameters and characteristics which define a system could be identically derived by any number of properly knowledgeable persons. It could happen, but probably never will. The complexity of system design factors and of their interactions and interdependencies means that searches for solutions involve many trade-off considerations which are highly probabilistic in nature. Also, the search for truth must be conducted through continuous and free discussion, open to the competition of all ideas, evidences, and arguments. These are tempered in a bureaucratic environment by consideration of factors concerning technical feasibility, reliability, maintainability, and safety, for example, but they are also biased by emotional criteria. The emotional environment stems in turn from ego involvement with organizational status, competition for power, and inherently by a prospective monetary reward as a result of impressing management with technical and management capability and soundness of judgment. This factor is compounded in the execution of complex system design by the difficulties in communicating through the complex human interrelationships in bureaucratic organizations both on the customer's side and on the side of the producing organization. It results in a climate for decision making which involves filtering information at all echelons in the hierarchy, both in transmission from the purely technical level upward to the purely political level, and downward from the latter to the former level. The catalysts at each level of filtration are the politics and the economics of the situation as well as the amount of talent and skill of the leadership of each level.

Since system design approaches are naturally probabilistic in terms of degree of confidence for predicting their ultimate performance success or failure and, therefore, permit flexibility in technical decision making, and since such decision making takes place in a relativistic political-economic-technical-human factors environment at all levels, *all system designs are fallible*. It is not possible to derive the perfect or even the optimum design. There will always be mistakes. The most effective system project management can only hope to achieve a higher percentage of good decisions than will those managements which are less effective.

The rest of this treatise will be devoted to discussing how the conditions for increasing the effectiveness of system design and engineering decisions can be improved. The underlying philosophy for doing so will be pragmatic, since there is no absolute way of ensuring that the humans who must exercise the decision-making authority can be standardized in the ways they will think or react to a given situation. Understanding of the basic scientific process and of the rules for effectively applying the process are essential prerequisites. It is man's capability for understanding the regulated process which makes the application of the scientific method to system design and engineering possible. Conversely, it is his inclination to ignore the process when exercising biases in decision making which makes attention to the scientific principles of system design and engineering necessary.

The Evil of Complexity

The most critical problem to be faced in modern large-scale complex systems is complexity itself. Unnecessary complexity is largely attributable to the bureaucratic approach to system project management. Large system project management organizations result in fractionating system design tasks and assigning them to numerous individual specialists. When this is done without first developing an overall system design as a firm framework, the bits and pieces are designed and developed as a way of accomplishing specific system functional input–transformation–output tasks. System engineering is employed only in the sense that adaptive devices must be designed and employed to tie the individual bits and pieces together in order that they can function as an integrated whole. The really challenging role for system engineering is to determine the design requirements for integration first and then to ask the specialists to supply the detailed technology for implementing the stated design approach. Only in this manner will we be able to invent the simplest solutions to complex problems. The remainder of this book is really concerned basically with the ways and means of attaining this fundamental objective through the application of an orderly system design and engineering process.

3

Organizing for Success

The Criteria for Effective System Management Organization

The major theme of this book is that effective system designs result only when there is talented and knowledgeable, *system-oriented* leadership. However, providing such leadership with an appropriate system project management organization will most certainly help realize its full potential. If you the reader still do not fully comprehend or understand the difference between system design and hardware design, perhaps if you persist through this chapter and the ones which follow, the importance and meaning of the distinction will become clearer. One of the ways to emphasize the difference is to describe the organization for effective system project management and its role in directing and controlling hardware design efforts. The criteria for such an organization are the following:

Facilitate Communications. As depicted in Figure 1, system designs are dependent upon the effective integration of multidisciplinary efforts. In most of the problems encountered in designing an organized man–machine system, each of the disciplines concerned can make a significant contribution on their own initiative. But, from a system standpoint, few of the problems that arise can be adequately handled within any one discipline. Any particular system design is not fundamentally electrical, electronic, mechanical, chemical, hydraulic, physiological, thermal, psychological, social, informational, economic, political, or ethical. Such disciplinary backgrounds merely provide different ways of looking at a prospective system design. Complete understanding of a proposed system design requires an integration of all perspectives which are involved. By integration, we do not mean a synthesis of the results obtained by specialists in the

21

various disciplines independently conducting separate system design studies, but rather of the results obtained by conducting a coherent system design study in which all of the required disciplinary perspectives have been synthesized into a single task force effort with a unitary purpose. The integration, therefore, must come during, not after, the performance of separate specialty-oriented analyses. The organization of a system project management office should provide the opportunity for all disciplinary specialists to work directly together continuously on a face-to-face basis and, most importantly, to acquire the systems viewpoint and an understanding of the role that their various specialized knowledges can play in deriving a particular system design. In fact, it must be stressed that a participant in an integrated system design effort is no longer working as a specialist but as a *generalist.* He must, therefore, be able to use a commonly understood system-oriented language, and not just his specialist-oriented jargon, which, when employed by any number of specialists in relation to a system-oriented context, can result only in a babel of tongues.

Streamline Controls. For efficient system project management it is essential that a clear delineation be established concerning the level of detail that will be controlled by a system project management office. Since, as has been pointed out, project management offices must of necessity be staffed by individuals who have "graduated" from a specialty, it is a natural tendency for them to slip into consideration of design details which are best delegated to subsystem or component level design managers. By simplifying the organizational structure and limiting the number of people who will be employed to staff the functions, this tendency can be discouraged, especially if it is properly coupled with appropriate leadership and job indoctrination.

Simplify the Paperwork. Most scientists and engineers for some reason naturally resist accepting a given form or procedure for writing and reporting. However, it is essential to standardize communication media for efficient system management. In selecting personnel to become system-level managers, it should be stressed to them that they are leaving the world where equipment or service end items are produced and are entering the world where paperwork is the only product. Over the period of years, specific means of communication have been developed by which system requirements analyses, trade-off studies, and system designs can be recorded and transmitted. At the system level these include such things as functional flow diagrams, schematic block diagrams, outboard and inboard profile drawings, facility layout drawings, mathematical and statistical formulas, and various technical description formats. However, it is an accepted engineering procedure that a system design is finalized and, therefore, formally described only in specifications. These range from the top system specification, through subsystem specifications, to end item specifications. The activity of a system project management office should be limited to producing only that level of performance/design requirements specification detail required to contract with the managers of end item design and engineers who, in turn, will prepare detail design specifications for producing and testing end items. Normally, therefore, system project management offices should be responsible only for preparation of system and subsystem performance/design requirements specifications.

Concurrent with the system design and engineering activities of a system project management office there is also the need for preparing the management plans for conducting the business aspects of budgeting and controlling the expenditure of funds allocated to accomplish the necessary complete system and end item analyses, design, engineering, fabrication, assembly, integration, test, deployment, and logistic support activities for the complete system life cycle.

Basic Model for System Project Management Organization

Figure 3 depicts a basic organizational structure for system project management. It has been drawn to illustrate the general pattern of those system project management organizations judged

Figure 3. (see foldout opposite) Basic system project management organization.

as having been reasonably successful in managing some major system design and development efforts. In relation to the criteria that we have been discussing, it is a *product-oriented* management structure aimed at providing the capability to carry out all system engineering and system management activities which are essential for designing and producing a *total* system to meet a customer's requirements. It should be assumed that both the customer and the producer—for example, a government research and development agency and an industrial contractor—should have the same basic system project management organizational structure in order to facilitate communications between them.

We will now briefly describe the role of each project office activity as shown in Figure 3.

System Project Manager.   This is the individual responsible for managing all activities concerned with planning and executing the system design and development project. Functionally, his responsibilities are those common to all executive positions; that is, planning, organizing, coordinating, controlling, and directing. From the standpoint of achieving effective system designs, it is particularly important that a system project manager be knowledgeable and able to ensure that all aspects of a system are accorded equal functional importance no matter how minor a role some particular specialized area may play in a given system design. If it is involved at all, it is important in terms of attaining ultimate system success. For creating complex system designs such as we are addressing ourselves to, it is impossible for any system project manager, no matter how intelligent, knowledgeable, or creative he may be, to know all the facts that he needs at any one time. He must, therefore, employ such management aids as he deems necessary to further his understanding and to support his decision-making authority. Without facts he will make all necessary decisions; with facts, he will make all necessary decisions, but with much greater understanding and accuracy. It is the purpose of a system project management organization to be able to keep a system project manager properly informed in such a manner that his capability for intelligent decision making is enhanced.

Associate Project Manager for System Engineering. This position, along with that of the associate project manager for program control, are incorporated within the system project manager's organizational block. These are the two principal functions of system project management. However, the scope of their combined activities is such that a system project manager could not be expected to be responsible for supervising the number of separate tasks to be accomplished. The associate project manager for system engineering should be responsible for establishing the system design and controlling its implementation and verification. Here are some of the functions normally under the direction of the associate project manager for system engineering.

System Design and Engineering. This office monitors and furnishes technical direction, via specifications and specification changes, for system and design engineering activities required to translate system performance requirements into subsystem (or major segment) design requirements. In turn, the various subsystem (or major segment) offices, as appropriate for a given system design and development effort, have the same relationship with end item (or subsystem) design management offices. The detailed design groups are the functional organizations within or without a system contractor's plant which design and develop the system equipment and service end items.

Interface Control. The design and development of large complex systems usually requires the use of a number of associate contractors or subcontractors specialized in producing selected subsystems or major equipment or service end items. Usually the group of contractors, or even of subdivisions of a prime system contractor, will be separated geographically. The problem of integrating the activities of these separate entities can be quite severe. An interface control office to provide for communication, coordination, direction, and control of design and development activities of geographically and administratively separate organizations contributing to a system design and development effort is essential.

System Engineering Analysis and Evaluation. This function is concerned with accomplishing the major tasks indicated on

Figure 3. Its most important role is to plan, guide, and document the application of the system engineering process as appropriate for a given design and development effort. It must also continually evaluate the progress in the design and test of equipment, software, and service end items in relation to their eventual integration into a coherent system entity capable of attaining the desired system performance effectiveness.

System Integration and Test. Planning, monitoring, and evaluating the results of an integrated system test program is another major activity under the direction of an associate manager for system engineering. For administrative convenience, this activity can be further subdivided into engineering test, qualification test, and system integration and operations test. These test functions are generally performed in different geographical locations and require different kinds of instrumentation and test personnel skills. However, they must be planned and executed as an integrated sequence of test operations. Finally, the results must also be evaluated in relation to achieving not only end item performance requirements, but for demonstrating attainment of system performance requirements when assembled and merged into a functionally complete system. A well-planned test program progressing from early experimental breadboard/brassboard equipment models, software simulation models, end item engineering and qualification models, and production prototypes of operationally configured end items is the most effective system engineering process because it best incorporates the step-by-step problem-solving approach to establishing a system design. The crux of the scientific method, as applied to system design and development, is a well-planned series of "experiments," or "prototypes" designed to progressively reduce the technological uncertainties inherent in the design of new and untried component end item designs, and ultimately of their integration into a unique system design. Planning and controlling the implementation of such a test program is the most effective means available to the associate project manager for system engineering for optimizing the performance effectiveness of a system design. System engineering analyses are important and significant for deriving an initial system design approach, but they cannot be substituted for the "cut-and-try" method of demonstrating an unverified technological approach before attempting to use it as a building block in a new system design.

Associate Project Manager for Program Control. This office is responsible for administering those "overview" activities which provide the indices of how well a system design and development effort is meeting the planned technical performance, cost, and schedule goals. Actual evaluation of technical performance is a system engineering function performed as an integral aspect of the system engineering evaluation and control task. However, the results of such a technical evaluation must be translated into system project management objectives and interrelated with cost and schedule considerations. The primary vehicle which is employed for exercising this control is the work breakdown structure and the resulting definition of tasks and their interdependencies on a time-based "network" graph spanning the system design and development effort to be undertaken. In addition to this key program control task, there also are the necessary tasks of ensuring that contract data requirements are properly compiled and delivered on schedule to the customer and that the administration and support of project activities—the "overhead" charges for accomplishing the technical task—are properly covered and are kept within the provisions of the system design and development contract.

Configuration Management. This activity establishes the uniform communication media which will be used by all other activities for identifying, recording, controlling, and accounting for the status of directive information which is used to regulate the process of system, subsystem and end item design, engineering, fabrication, assembly, integration, testing, and eventual deployment, logistics support, and reprocurement. It in fact constitutes the system project manager's secretariat by which he keeps informed and is able to control and maintain an integrated and coherent system design and engineering effort. He will be aided in making decisions in terms of obtaining technical and management analyses and information by dealing directly with the other activities shown on the organization chart, but their recommendations and his decisions will have no authority until recorded and transmitted through the configuration management specification program. Funding of end item activities and authorization for spending must be contingent upon issuance of system project manager approved specifications or specification changes (which may carry additional funding).

Quality Assurance.   This is the classical function of establishing and administering an inspection procedure which ensures that all contractual requirements have been fulfilled upon delivery of the products of a system design and development effort.

Contract Management.   This function is also a classical one and is concerned with the legal aspects of fulfilling the terms of a system design and development contract.

Customer Liaison Office.   A customer's understanding and acceptance of the end products of a system design and development contract is facilitated by having his technical representatives quartered with a contractor's system project management staff. Their participation in technical discussions and continuous reviews of the design process within the bounds of contractual provisions on a "noninterference" basis helps in developing an environment favorable to mutual understanding between the contractor and his customer of the design and development effort. It furnishes a basis for expediting the contract changes as may be required to incorporate design changes resulting from progressively more detailed engineering analyses and tests. Since customer representatives cannot be directly controlled or supervised by a system project manager, their relationship is shown by a dash line on the organization chart depicted in Figure 3.

Production Management.   Finally, completing a description of the basic functions that should be covered in a system project management organization, is the important function of planning the production of the equipment end items. There are two major tasks to be accomplished, namely, how the end items and all of their myriad parts, materials, and processes are to be procured, and how to build up an item when all of its parts are assembled.

The process of buying and manufacturing an end item is, of course, a textbook subject in itself and is certainly well beyond the scope of this book. It is included here merely to complete the picture of a basic system project management organization and to indicate the interface of that function with system engineering via the system project manager's office. The requirements for fabricating a given system or end item design can most certainly be an important constraint to be considered in system engineering analyses aimed at deriving an optimum but cost-effective design approach.

Again, it is emphasized that it is not the purpose of this discussion to provide detailed system engineering and system management operating procedures, but only to furnish a broadbrush description of the basic approach to system project management which will facilitate, rather than hinder, the attainment of effective system design and engineering practices. Specific system design and development efforts will require tailoring a project office organization to meet specific needs, but any such organization can be built upon the basic structure that we have been describing.

Finally, centralization of decision-making authority is a way of life in the management of complex and large-scale systems. Simplifying such centralization and limiting decision making at all levels in the system hierarchy to that level of detail necessary to enable each successive lower level to properly undertake its work are essential. Such an approach, when properly defined and regulated, permits visibility and traceability of system design and engineering requirements throughout all echelons of system project management.

Work Breakdown Structure

To determine the specific project management and system engineering organization required for a given system design and development effort, the preparation of a work breakdown structure is a most useful tool. The steps in deriving a project work breakdown structure are as follows:

1. The scope of the work to be accomplished must be analyzed and accurately described, including the project limits and constraints.
2. Functions and tasks required to perform the work are then identified and defined.
3. A logical grouping of the functions and tasks are depicted graphically as a work breakdown structure tree as outlined in Figure 4.
4. A work package dictionary is prepared to describe the specific detailed work to be accomplished to complete each task, including inputs from and outputs to other project tasks.
5. Finally, the interrelationships and relative time-phasing of the tasks are depicted on the network diagram outlined in Figure 4.

This last step will permit establishment of calendar dates when each task is to be initiated and completed within the overall framework of the project milestone schedule. This milestone schedule should be drawn to depict completion dates of events required to achieve project "go/no-go" decision points. For example, such a schedule would depict completion dates for such breadboard/brassboard engineering development tests as may be required to conduct a system preliminary design review; completion of qualification tests prior to a critical design review; and so on to project completion. Each scheduled design review should be employed as the basis for deciding whether analyses and tests completed to the date of the review indicate satisfactory progress in demonstrating fulfillment of system performance requirements. If so, proceeding to the next planned phase of design and development would be authorized. If not, then the design and development tasks should be modified and the milestone schedule revised accordingly.

The project task network based upon the work breakdown structure, in addition to providing a means for tying the tasks to the project milestone schedule, will also identify the lines of communications among tasks in terms of their interdependencies and mutual constraints. Thus, it will furnish management visibility for structuring the work effort by depicting when each task should be properly and most effectively accomplished.

When a project work breakdown structure is used to define and organize a given system design and development effort it will be revealed that different kinds of tasks and personnel skills are required for different phases in the life cycle of a project. Furthermore, it enables project management to properly tailor the project organizational structure for acquiring and properly utilizing the proper mix of generalist and specialist skills in relation to the nature of the tasks and work to be performed for a given phase of system development.

Adopting this method for organizing a project management effort virtually assures a product-oriented system project office organization.

Historical Note on System Project Management

Readers who are familiar with system management practices in the approximately two decades since the "system engineering" approach was first developed at Bell Laboratories (first publicized at MIT in 1950) will immediately recognize that the organizational approach discussed in this chapter has probably never really been fully implemented for a major large-scale man–machine system development project. Generally speaking, system project management has been *function-oriented*, rather than *product-oriented* as we have described it. By this it is meant that under a system project manager there are parallel subproject management organizations devoted to such "functional" efforts as prime equipment design and development, support equipment, assembly and test, reliability and maintainability, and various specialized technologies. At the best, system design and engineering principles have been applied most successfully to major system segments or subsystems, particularly in relation to support equipment design and development.

In relation to a total system design and development project, system engineering has more typically been included in the

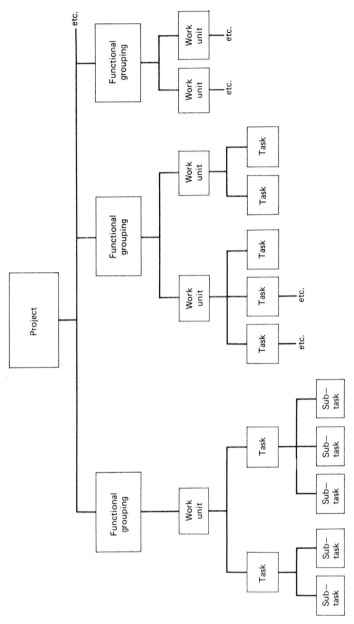

Figure 4. Outlines for a project work breakdown structure tree and network.

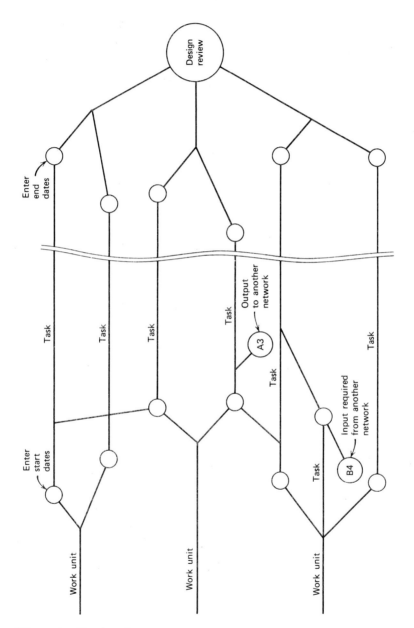

Figure 4. *Continued*

system project management organization as one of the parallel functional activities rather than as an overall management function for deriving a unitary coherent design of all system segments, subsystems, and components. As a consequence, system engineering has largely been an exercise in producing massive volumes of descriptive material after-the-fact of equipment design. System engineering has been viewed as being virtually synonymous with preparation of such documentation and as such has been in disrepute among hardware-oriented engineers. Often there has existed a "talent inversion" in project management organizations in respect to positions occupied and level of responsibility assigned to personnel proficient in system design and engineering knowledge and skills. Those most qualified in system methodology and techniques have generally occupied lesser management positions and have exercised little influence upon the derivation of system designs. Key system design decision-making management roles have generally been assigned to those who are best qualified for prime equipment design and engineering. This has meant that there has been an admixture of management interests within system project management offices, with only minor emphasis on system-level considerations. Under such a regime the main management interests have been concentrated on directing and controlling design details which are best left to subsystem, end item, and component designers, provided they are given proper technical direction and specifications in regard to the performance and design characteristics which must be achieved in order that detailed designs will properly integrate to provide the desired unitary functional system performance capabilities.

Advantages of the Product-Oriented System Project Office Organization

Since complicated systems, especially large-scale ones, require the use of many different technologies and design disciplines, the use of a bureaucratic approach in system project management organization cannot be avoided. However, it is possible to avoid the "bureaucratization" of the design approach. By streamlining the organization and integrating the various specialist backgrounds into common system-oriented task groupings, the loyalties of individuals can be directed to the systems effort. It is not unusual in a function-oriented system project management organization to hear a participant in a discussion of a given design approach say, "My boss told me to take this position. However, from the system viewpoint. . . ."

"Bureaucratization" of system project management frequently results in the adoption of some technological approaches which from the systems viewpoint are inappropriate solutions to given problems. Effective team work with a common purpose can be more easily achieved in the product-oriented system project management organization. The function-oriented system project management approach sometimes encourages a manipulative, often furtive style of leadership concerned more with "feathering one's own nest" than with achieving an integrated system design and engineering effort. It contributes to "bureaucratization" of interests because there are many bosses whose continuing job existence is dependent upon maintaining a significant organizational role.

From a systems viewpoint, compartmentalization of design efforts makes lateral communication very difficult. Unrestrained, the functional groups proliferate design approaches unilaterally to suit their respective interests. Elitism is encouraged. As system design and development proceed under this kind of system project management approach, belated system engineering effort must be devoted to "crisis management," in order to achieve integration of separate components and to make them "play together." A rigorously enforced system design approach under the control of a relatively small system project management organization should reduce the need for extensive "make-it-work" efforts.

System-oriented leadership will always be in short supply. It is essential that it be conserved and provided with an organizational structure with clearly identified and easily understood lines of authority. Any temptation to establish informal lines of communication and authority must be discouraged. An intelligently directed product-oriented system project management office creates a situation where properly qualified system-oriented personnel can experience success in deriving an effective system design. Finally, it provides a basis for motivating potentially good system project managers to accept responsible positions at the management level.

4

Absorbing the Cults

Parallelism in Functionally Independent Design Activities

In function-oriented system project management organizations where interest is mainly centered on detailed designs of subsystems and components, and with a weak overall system engineering effort, it has only been natural that a number of specialty groups interested in improving and controlling the quality of the system product for a given design characteristic have tried to simulate a system approach. Since the interest that each represents constitutes only one aspect of the overall system design and engineering effort, and since in function-oriented system project management offices these aspects have been organized as separate parallel functional activities, their efforts at best have had a limited impact upon achieving performance-effective system designs.

The full realization of the objectives of these several system-oriented design activities is dependent upon a strong central product-oriented system design and engineering management. A brief description of the objectives of these design activities will indicate why this is so. They employ common techniques and have redundant system interests in pursuing their objectives which collectively are the techniques required for accomplishing integrated system design and engineering. We will, therefore, be describing in more detail those basic activities which must be the direct responsibility of the system engineering activity of a system management project office as shown in Figure 3. Generally speaking, the hardware-oriented specialists who are involved in the mainstream subsystem and component design effort have come to refer disparagingly to these activities as "cults."

System Requirements Analysis and Integrated Engineering Data Management

Assuming that, as shown in Figure 1, operations research and technical feasibility studies have established the need for a system and have set forth the basic requirements to be met in its development, the next step in relation to an overall system design and development project, and the first step in the system design and engineering effort (concurrent with undertaking system design studies and the establishment of system performance-effectiveness criteria as shown in Figure 2) is to define functional requirements to produce and use a system. Normally, this effort will be initiated with the preparation of scenarios setting forth descriptions of how it is conceived that the proposed system will be operated and supported in order to accomplish its mission objectives. When these are confirmed in relation to the customer's stated system requirements, if any, operational, maintenance, and logistics support criteria can be prepared. Subsequently these are translated into progressively more detailed operation and support functional sequences. Sequence charts or flow diagrams and functional requirements descriptions are prepared. These are collated with design studies setting forth the possible technological approaches which can be employed in the system design. Also, the operations and support requirements are studied in relation to the system performance-effectiveness criteria as a basis for deriving an optimum design approach for implementing the integrated system requirements in a manner which is best for achieving the system mission within the constraints of the customer's stated cost and schedule limitations.

Ultimately, through the process of iteration, a complete set of performance requirements will be compiled. These are employed as a common data base in specifying design requirements in regard to the performance characteristics of the prime equipment, supporting equipment, facilities, computer programs, procedures, personnel training, and logistics support. The system project management office should conduct the requirements analysis only to the level needed to prepare the specifications required to contract with or to direct the next lower level of activity to proceed with determining the performance requirements for designs of subsystems, end items, and their components.

A complete program of functional requirements analyses will encompass all aspects of system end item design and development. Progressively, these analyses will cover: (a) mission operations; (b) equipment servicing and maintenance; (c) facilities; (d) support equipment; (e) procedures (including any computer programs) for operations, maintenance, and support; (f) personnel selection and training; (g) production, assembly, integration, test, and deployment of end items; and (h) logistics support of the operational system.

An integrated engineering data dissemination program should be employed under the control of configuration management. This program should communicate system and end item design requirements via a uniform specification program between all levels of design activity from the system level through the subsystem, component assembly, subassembly and down to the irreducible level of detailed parts, materials, and processes. From a systems standpoint, a product-oriented system project management office should manage only the process of implementing and verifying the attainment of the uniform specification process at all levels of design and development management. Otherwise, it will be committing the typical error of function-oriented project management and be engaging in making detailed design decisions. When this happens a system project management office tends to lose the advantage of maintaining an impartial objectivity for evaluating the adequacy of proposed detailed designs from the standpoint of how well they meet *total* system performance requirements.

In practice, formalized engineering data management has involved developing intricate reporting schemes which attempt to document infinite details. The justification has been that formalized and exhaustive documentation of design details for complete visibility and traceability is required for system engineering management purposes. However, such an approach cannot foster or stimulate perceptive system management leadership, but it does result in massive and, therefore, little used documentation efforts on large-scale system design and development projects.

System Performance Effectiveness

As an integral aspect of the system analysis process (see Figure 2) some measure of performance effectiveness is required as a basis for evaluating proposed design approaches. This can range all the way from informal, unstructured qualitative value concepts employed by the individual system analysts, designers, and managers to a highly structured quantitative mathematical model which can be employed only by a specially trained system analysis expert. In either case the application of system performance-effectiveness criteria to evaluate design approaches, whether qualitative or quantitative, is always finally a matter of individual judgment.

In the more structured approach to establishing criteria or standards, system performance effectiveness is an integration of system performance capability and its availability. In turn, the latter is an integration of reliability, maintainability, and, when necessary, of survivability/vulnerability (see Figure 10). When systematically applied, it is necessary to establish a means for collecting and accumulating test and operational performance data which can be employed to verify the degree to which a given system design attains the established system performance-effectiveness criteria or standards. This process extends over a considerable period of time to include system performance testing under environmental use conditions. The empirical results obviously cannot influence the original design of a system, but they can lead to modification of the design or to improving the designs of similar follow-on systems. Successful use of the approach requires a continuity in overall management in the design and development of a family of systems so that experience gained in one system project can be applied directly to subsequent ones having similar characteristics.

Design and Trade-off Studies

As shown in Figures 1 and 2, design and trade-off studies constitute the mainstream of the system design effort. When this effort is properly supported by functional requirements and system performance-effectiveness analyses, it is possible to conduct trade-off studies which will include consideration of operational use objectives as well as technological ones in alternative system design approaches. In carrying out this system-oriented design approach, all of the so-called "ilities" which have been nurtured by the "cults" will be appropriately applied as integral factors in trade-off studies for deriving system designs. These include *reliability, maintainability, safety engineering, human engineering,* and *value engineering.* The phenomenon of specialists in these areas trying to use their respective disciplines as forcing functions to obtain some system considerations in equipment design will disappear. Each area will be considered on its own merits in the normal course of events in deriving a given system design. A brief definition of each follows:

- *Reliability* is the probability that a system will perform correctly for a specified time interval under a set of specified conditions.
- *Maintainability* is the characteristic of the system equipment that permits ease and economy in accomplishing all maintenance functions. This includes such things as ease of servicing and of removal and replacement of components, rapid fault isolation, failure prediction, minimum personnel skill levels for maintenance tasks, and so on.
- *Safety Engineering* is concerned with system design characteristics which minimize the chances for unsafe operating conditions and for inadvertent damage or destruction of equipment or personnel.
- *Human Engineering* is applied in system design to obtain equipment characteristics which facilitate the error-free and safe performance of operation, control, and maintenance tasks.
- *Value Engineering* is oriented toward eliminating unnecessary system features and reducing overstated requirements, thereby lowering costs without impairing operational capability.

Mathematical Modeling

The type of complex systems that we are discussing incorporates a bewildering variety of discrete hardware and human functional components in order to accomplish defined system output characteristics. An important part of the system analysis effort is to identify and describe the appropriate *deterministic* and *probabilistic* relationships which predict the interactions among the components (both human and equipment). When these functionally multivariate interrelationships are properly understood, it becomes possible to construct mathematical models which can be programmed for computer simulations for studying various performance characteristics that will result from given combinations of input–transformation–output values. Whether such simulations will be accomplished on the *total* system level and will affect overall system design approaches depend upon the technical skill, understanding, and leadership of system project management for properly using them. In the product-oriented system project management office, mathematical modeling would be used for furnishing comparative data needed to study the results that can be achieved from alternative functional relationships. In the function-oriented organization of system project management offices, like other potentially useful system-oriented techniques, mathematical modeling is likely to be relegated to an incidental role, and may, therefore, be looked upon only as an interesting toy. As such it is likely to be either ignored if it suggests results which are incompatible with favored design approaches or praised if it confirms their adequacy. In the latter case, it more often than not furnishes only "gold plating" for what already has been decided on some other basis.

Interface Control Documentation

In a function-oriented system project management organization where interest has been mainly centered on detailed design of subsystems and their end item components, and where there has been a weak overall system engineering effort, it has been necessary to prepare special interface control procedures and documentation. In fact, in such a situation this effort has been synonymous with determining system integration and test requirements in order to properly incorporate separately produced components into higher order subsystem or system functional entities. The need for interface control documentation largely disappears in a product-oriented system design and engineering effort such as we have been describing. Subsystem and component input–transformation–output interface information is built directly into the system requirements analysis and data recording process which is used as a common data base for preparing all end item performance/design requirements specifications.

It may be that for really large-scale systems which are an aggregate of individually complex but interrelated special purpose subsystems, manual compilation and collation of system functional requirements analysis data may be impractical. If so, it will be necessary to employ computerized information processing to ensure that all interface functional requirements among the various systems are properly identified and interrelated. The compilation should furnish the basis for deriving the performance and configuration design requirements to be specified for the design of individual subsystems to maximize their characteristics in terms of their physical and functional compatibility for joint installation and operational utilization with respect to one another, as well as with respect to their common use environment.

Integrated Logistics Support

Operational deployment of a newly developed system brings its own headaches. This is when its true *supportability* characteristics really become self-evident. In an attempt to feedback field experience to designers, the logistics support specialists document their analyses of system and equipment design inadequacies as they show up in system operational use. The accumulated experience results in maintainability design criteria. In addition to maintenance and repair considerations, such criteria take into account highly correlated reliability, safety, human engineering, and value engineering factors.

Logistics support experience on consumption rates for certain types of components, parts, and materials in given operational and environmental use situations can be very valuable in designing new equipment intended for use in identical or similar situations. Further, such information helps in estimating life cycle costs for new systems and, therefore, furnishes important data for operations research analyses concerned with estimating the cost effectiveness of alternative proposed system designs. Obviously, a system which will employ rugged, long life, highly reliable, nonreparable, easily replaced, cheaply produced components is going to be more cost effective than one which does not possess such characteristics, provided it can be employed just as effectively to accomplish the defined mission. Sometimes to accomplish a given mission of great importance a high value item is required, and there is really no "cost-effective" way otherwise to attain the desired performance capability.

Life-Cycle Costs

An integral aspect of cost-effectiveness analyses of a candidate new system is consideration of its potential life-cycle costs. Once a decision has been made that the investment in new technology and system development costs will produce a needed performance capability, then consideration must be given to what other costs will be incurred in acquiring the new system. Here are some basic parameters involved in analyzing and estimating the probable "costs of ownership" of a new system.

Facilities. Will new or modified facilities be required to shelter, operate, control, maintain, and/or support the system? Facility requirements for a new system can easily pyramid, especially if they incorporate significant increases in the number required, their size, and/or complexity. For example, the new system may require unique facility characteristics for usable space, access, structural strength, environment control, and perhaps special power, plumbing, lighting, sound proofing, safety provisions, and the like. Old facilities, therefore, may not be easily or economically adapted to be compatible with the new system.

Personnel Skills. Even though a new system may incorporate the best attainable designs for personnel operation and maintenance, it could also possibly require an increase in personnel skills to perform operation, control, and/or maintenance tasks. For example, if there is an increase in system complexity, maintenance activities could involve an increase in skills needed for repair of components, unless, of course, it is economical to go the route of "throw away" of failed components. To make this decision, the expense of procuring and stocking replaceable components must be traded-off with the cost of reparable ones, including the associated cost of training skilled personnel to repair them. Performance reliability of the system components under consideration is also an important factor in determining the relative cost-effectiveness of replaceable versus reparable components, including impact upon needed personnel skills. Of course, increased operator skills may also be required for a candidate system design, and possibly even new management skills. In other words, increased personnel skill demands of new system may ripple throughout a whole organization dedicated to its employment and support. Higher skills normally increase system operational costs, both in terms of training costs and salaries to attract and retain qualified personnel.

Maintenance Support Equipment. Will the new system require development and/or acquisition of new support equipment for the prime mission equipment? These costs can be considerable if special mechanical handling gear and/or electrical/electronic test and servicing equipment must be acquired to achieve a low mean-time-to-repair capability in order to achieve the desired operational "up time" for the prime mission equipment. Support equipment in its turn also requires proper facilities, personnel skills, tools, test equipment, and consumable supplies to keep it operational. There are plenty of examples of systems being nonoperational because the proper support equipment was not available to keep the system in an operational ready condition.

Failure to plan for and to provide appropriate support equipment for a new prime equipment system is not cost-effective in the long run in relation to realistically estimating potential system life-cycle costs.

Consumables. All systems must be fed in order to operate. They all eat up consumable supplies at some rate, and often these rates are fantastic and voracious. All systems require energy, either electrical or combustible. Lubricants (sometimes exotic) are generally needed. Most systems require printed paper forms, and computer systems especially consume great quantities of printed paper forms and also magnetic tapes. Some systems even require supplies of special gases and fluids for such things as cleaning, preservation, corrosion control, and cryogenic cooling. To properly derive life-cycle costs, the cost of supplying needed consumables, including any inflationary factor over the expected life of a system, must be expertly estimated.

Computer Programs. Computer driven or controlled systems require software maintenance and updating to keep computer programs current with system operational, control, or maintenance requirements. Programming costs are not inconsiderable over the period of time that a system is expected to be in use. They are an important element in determining potential life-cycle costs of a system's computer(s).

Spares and Supply Support. To acquire and supply a system with spares and consumables requires warehouses, transportation, inventory control, and personnel, as well as manufacturing facilities and production lines. Acquiring a new system without considering the costs for feeding it what it must have to operate effectively can be an important, and perhaps a catastrophic blind spot when it comes to accounting for unpredicted increases in costs of system acquisition and ownership.

Obsolescence and Phase-out. All systems, like life itself, have some rate of degeneration and consequent degradation in performance capability. Or, new technology developments can artificially cause a system to become obsolescent before its anticipated wear-out rate would normally render it too ineffective, inefficient, and uneconomical to use. The longer a system can be projected as being operationally useful without becoming uneconomical to maintain or technologically obsolescent, the more justifiable will be its initial cost of development and acquisition. Sometimes, the costs of phase-out and disposal of a system are excessive, especially if it has not been possible to amortize initial development costs over a considerable period of time of operational use. An approach to reducing the threat of too early obsolescence for a system is *modular design*. Various components of a system are designed and installed with the view that they could be replaced on an individual basis if it becomes desirable to technologically upgrade the function they serve in order to improve overall system performance. Such an approach precludes the necessity to design a whole new system in order to gain an improvement in the performance of a given component or subsystem. This approach to system design could be particularly cost-effective when there are subsystems for which significant technological advances can be reasonably foreseen during the predicted useful life of the system.

There can be factors other than those we have discussed here which will be significant in estimating a system's life-cycle costs. Whatever the contributing factors are, differences in potential life-cycle costs must be an important consideration in the system engineering process for deciding between alternative system design approaches in order to select that approach which promises to be acceptably effective, efficient and economical to own. When potential life-cycle costs are taken into consideration in evaluating proposed systems designs, very often design concepts can be modified to effectively reduce them in some one candidate system design with insignificant or no losses in system performance capability, reliability, and maintainability.

Considering the ever increasing sophistication, complexity and costs of system development, acquisition, operation, management, and logistics support, a system engineering analysis must develop such life-cycle cost data as may be necessary to permit the prospective customer to satisfactorily answer for himself the question,

"If I acquire the system, can I afford to own it?" Or, conversely, "Considering the costs of acquisition and ownership and what the system will enable me to do, can I afford not to have it?"

Created Obstacles

The sincere efforts to devise means to counteract the ineffectualness of function-oriented system project management for accomplishing extremely complex system design and development efforts have led unwittingly to the introduction of obstacles to realizing the fundamental purpose for which they are proposed, namely, the attainment of a system engineering approach which is understood and effectively utilized by all participating groups. The various schemes which have been instituted must be reviewed and modified if a combined hardware designer and system-oriented engineering working-team relationship is to become a reality.

A major obstacle to such an effective liaison is the overelaboration of system management and system engineering schemes as salable products. This has progressed to the point that creative hardware designers generally refuse to learn what appears to them to be a wholly unnecessary mishmash of data documentation which is only artificially related to the deliverable working end product of their efforts. System project managers also tend to consider the resultant "system engineering management requirements" as being things that a customer may request, but which are not essential to the design process. As a consequence, the system-oriented engineer tends to gravitate to organizational groupings composed of like-minded individuals. The customer, especially when it is a government agency, usually supports these groups by contracting for various system engineering analyses and reports. However, groups devoted to such activities in function-oriented system project management offices normally have not been allowed to function as an integral aspect of the mainstream of system design. At best, they may obtain reluctant compliance with a few limited-impact design concepts.

A second obstacle that has worked quite disadvantageously

from an integrated system design viewpoint stems from the contemporary concern with one's "image" in reference to one's speciality group acceptance. Evaluating one's group standing encourages the organizational man against the independent-thinking man and can result in the subjugation of objectively seeking best-fit solutions to system design problems in deference to a special professional interest, or the "cultist" viewpoint. Furthermore, government contracting for system development programs supports various discontinuous system-oriented activities, such as reliability, maintainability, safety, and human and value engineering, by providing separate funding for each. The resultant special interest groups, each with redundant system engineering objectives, have become embedded in a bureaucratic organization. Consequently, they tend to become highly resistant to any mutually supporting teamwork approach such as is essential for obtaining effective system designs. The isolationist tendencies of such groups within a presumably common system design and development framework are further compounded by a language translation problem. Each special activity group, whether it is system or hardware design-oriented, converses in its own dialect used primarily to enhance its own self-generated elitism. This difference in tongues has immeasurably slowed the development of universally accepted system engineering practices that can be commonly understood by both hardware designers and system-oriented engineers. Meanwhile, each of the system-oriented special interest groups stimulate the proliferation of papers and symposia intended to promote the improvement of system design. Most of the presentations are made from the viewpoint of furthering the role of a particular group as being the leading force for such improvement. At best, the techniques advocated by such groups are only partial, and often they are superficial, approaches to the attainment of performance-effective system designs.

As a combined outcome of the above obstacles, those who become concerned with applying only the veneer of system engineering, without a demonstrated capability in applying one of the hard-core scientific or engineering specialties to the solution of system design problems, cannot hope to achieve system management responsibility—nor should they. System engineering schemes, like all management plans, are concerned primarily with the invention and elaboration of formalized techniques and procedures aimed at enhancing the role of some bureaucratic organization. Those individuals whose demonstrated technical proficiency in system design and development projects rests solely in furthering such efforts are destined to occupy only staff positions, regardless of any personal or organizational group aspiration toward attaining more significant in-line system management roles with decision-making authority.

As a corollary to the above outcome, there is a serious lag in providing properly for the requisite education and job experience to prepare hard-core scientific and engineering specialists able to accept and employ good system design practices, and ultimately to undertake comprehensive system engineering and system management roles. The practice has been to promote research scientists and hardware engineers to system management positions without their demonstrating a creative ability for selecting and integrating those factors which are most pertinent for satisfying system performance requirements when several highly complex and sophisticated technologies must be combined in order to achieve a functionally coherent and unitary outcome. This neglect in providing for the proper development of truly skillful system managers and system engineers is at least partially a consequence of the general impression engendered by outpourings of the system-oriented special interest groups to the effect that better system designs are resulting from the application of the particular system engineering practices they advocate. Their persuasion has tended to divert attention away from the need to provide ways for hard-core scientific and engineering specialists to become competent system engineering generalists as a fundamental qualification for becoming system project managers.

The particular techniques which are employed by the professional system-oriented specialist groups do have a beneficial effect when seriously applied. However, truly significant advances in system design can be achieved only when the total fund of available system engineering methodology is selectively and effectively employed by knowledgeable and capable design personnel at all levels of decision-making authority, from the system project manager down to the individual hardware designers. In other words, to achieve their intended purpose, the system-oriented methodologies and techniques of the "cults" must be absorbed and become accepted and used as an integral aspect of a unified system design and engineering effort.

5

Building Operational Considerations into Initial System Designs

Identifying System Functional Units

The transition from understanding the approach to hardware-oriented designs to understanding that for system-oriented ones requires the recognition that all equipment intended for system application is being designed to be used and supported by a customer with a set of needs and values by the satisfaction of which he will judge the utility of the end product. Consequently, the system designer must be responsible for the integrated design of all elements of a system in relation to meeting the customer's specified operational use requirements, whether they be explicit or implicit.

Figure 3 under the heading, "System Design and Engineering," shows the basic work breakdown structure for designing a *total* system. Since system project managers normally "graduate" from hardware design activities, especially from groups dealing with prime equipment designs, the most difficult change they must make is to accept the fact that design activities other than those directly connected with prime equipment design are, from a system operation and support standpoint, *functionally equivalent.* Overemphasis or underemphasis on any particular aspect of a system during the design phase leads only to unnecessary delays in its development and ultimate operational deployment. From a system standpoint, the design of any of its elements, including prime mission equipment, is interdependent with the design of the other elements, and subsequently the design of a subsystem or a component end item cannot proceed without consideration of the design of other subsystems and component end items. Individual systems will vary in their specific operational requirements, but for all systems the necessity for planned integration of the design of all basic subsystems and component end items is clearly apparent if a coherent system design is to be achieved.

Along with the concept that each system has a definable integrity, it is also essential to accept the definition of a *basic conceptual unit of a system as being the "input–transformation–output function,"* which can be depicted as follows:

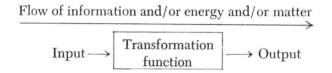

From a system standpoint all system elements, including the prime mission and support equipments, facilities, procedures, personnel training, and logistics support, must be designed as a coherent whole in order to permit carrying out integrated sequences of discrete input–transformation–output functions as required to produce desired performance capabilities.

In reading the descriptions of the steps in the system design and engineering process which follow, it is strongly suggested that frequent reference be made to Figure 1 for a description of the interrelationships of the various steps of the system design and engineering process and to Figure 2 for a description of the overall system analysis and design decision-making process.

Defining System Functional Requirements

The process which leads to the definition of system performance requirements starts with the translation of system mission requirements into basic functional sequences of operations to describe how the mission will be accomplished. It is a misconception that "pure" system operational requirements are analyzed and as a consequence "pure" functional activities are conceived and described, and then, and only then, are the ways of implementing these functions in system and equipment end item designs decided. It is a basic psychological fact that thinking never occurs in a vacuum. A given mission and the capability to accomplish it would not be under consideration unless there were some idea about the design of the prime mission equipment which could be developed to achieve the required operational objectives. Accordingly, functional analysis of system performance requirements always is made in reference to some concept of the nature of the prime mission equipment, supporting equipment, facilities, procedures, personnel, and logistics support resources that will be employed.

In other words, in conceptual design studies, equipment, facilities, procedures, personnel, and logistics support design approach concepts should have been thoroughly considered and evaluated for establishing system feasibility. By the time specific operational characteristics have been formulated for a system, quite firm design approaches, including use of available scientific and technological knowledges and skills should have already been decided. Realistically, therefore, the determination of *total* system requirements begins with identifying and analyzing the functions required to operate, control, maintain, support, produce, assemble, integrate, test, and deploy the defined prime mission equipment as a basis for determining the performance and design requirements for supporting equipment, facilities, procedures, personnel skills and knowledge, and logistics support. Gross concepts of these requirements formulated during conceptual design studies and any specific direction to incorporate existing end item designs as employed in other system applications are evaluated. If their adequacy cannot be confirmed, recommendations must be made concerning the need for their modification or for elimination of their use. Since normally the detailed design of the prime mission equipment will not have been completed, its design characteristics can be influenced for facilitating its operation, control, and maintenance.

The analysis progresses through more detailed functional requirements until all operation, control, maintenance, support, production, assembly, integration, test, and deployment requirements are identified and the complete performance requirements for integrated design of the system elements are decided.

Without the discipline of exactly defining the functional sequencing of input–transformation–output requirements to account for the total integration of all system and subsystem performance requirements, each designer called upon to translate a portion of the performance requirements into some particular end item design requirements must assume a set of input–transformation–output requirements in order to seek design solutions. The assumptions he makes may be grossly inaccurate in relation to tying his end item into the other components so that they can function as an integrated system.

Use of Functional Sequence Diagrams

Sequences of functions (serial, parallel, and floating) required to fully operate, control, maintain, support, produce, assemble, integrate, test, and deploy a system in order to accomplish its mission are depicted in "information flow" diagrams. The resultant set of functional flow diagrams is often referred to as the "functional model" of the system. In the preparation of such a model, it is incumbent upon each system engineer to be concerned with keeping a *total* system viewpoint, even though he may have a specialized interest in some subsystem. On the other hand, if he is responsible for representing a subsystem area, he must participate directly in the system-level analysis in order to ensure that major functions which may fall entirely within the scope of his subsystem interest in relation to the system design process are included in the relevant functional sequences of events and are properly depicted.

Identification of Performance Requirements

How each system-level function is to be accomplished must be broken down into qualitative and quantitative performance requirements to be levied upon each system element (or subsystem and component end item when carried to lower levels of detail). The gross functional operation requirements must be subdivided into subfunctions, the subfunctions into specific tasks, until a complete system functional requirement model is developed. Each resultant functional step requires a description of the input stimuli or signals and their values, and of the transformation to be made to produce the required output values. The importance or criticality of each function or task to the fulfillment of the system performance or mission requirements must also be established.

Description of Functional Characteristics as an Integrated System Design

The vehicle which is employed to depict the integrated system design is a series of *functional schematics*. These show in progressive detail the input–transformation–output interrelationships of all system elements, subsystems, and component end items. Generally they have been equipment-oriented. However, there is no real reason why input–transformation–output interrelationships between equipment and human components should not also be shown in these drawings.

Criteria and Standards for Evaluation of System and End Item Performance

Operational functions (including any maintenance functions which may occur normally in a given operation and control sequence) are *time-lined* in order to depict the sequential and parallel accomplishment of tasks as related to the time required to initiate and complete a specific system application or mission accomplishment. This constitutes the system performance standard to be employed in the system test and evaluation program as a basis for verifying that the design fulfills the stated system operational performance requirements. It represents the complete dynamic functional integration of all system elements working together to accomplish system use or mission objectives. As such it is the common performance standard by which all system elements must be designed and tested; otherwise there is no system. Sometimes, the time-line description of the system operation has been referred to as the "master plan" or the "script" for directing the roles the various system elements must perform on given cues as the operational mission unfolds. Attempts to proceed with the detailed design of system elements, subsystems, and component end items without such a master plan results only in chaos when the attempt is made to collect all the pieces together to perform specified system use applications or missions.

The time-line analysis (or the detailed operations or mission profile, as it is often called) furnishes the basis for identifying the significant or critical functions and tasks which will carry the burden for successful system performance. One backs off from these functions and tasks to identify the man–machine input–transformation–output elements that must function reliably within prescribed tolerance limits and within the allotted time period to achieve successful system use or mission completion. These are the functionally active elements of equipment and human performance to which allocation of reliability values must be made for achievement of the desired or specified overall *system reliability*. All other functions and tasks are supportive to the normally relatively small number of significant and operationally critical input–transformation–output functions that appear on the time-line analysis graph.

Interrelationships of Functional Requirements with Performance Allocations

A major concern in system design and engineering is accounting for the supply and demand interrelationships between functional requirements (demands) and the performance capabilities of the subsystems or subsystem end items to fulfill them (supply). The interrelationships can be shown in a double entry table, or matrix. As each function required to fulfill a system objective or to accomplish a specified mission is analyzed, the demands it places upon each subsystem or end item performance capability as its role unfolds on the time-line analysis are entered as quantitative values in the appropriate cell in the matrix table. (See Figure 12, for an example of a matrix presentation.) Some functions will impose a constant uniform demand upon a given subsystem or end item. Others will present variable demands. In the latter case, it is essential to identify on the time-line mission profile where the peak demands or worst cases occur. There will be instances where such peak demands combined with the demands that other functions may make upon the subsystem or end item will result in an overload condition. Overloading of any given subsystem or end item performance output capability at any point on the time-line analysis necessitates a reevaluation of system design to ascertain whether a functional requirement can be modified to reduce the demand, or whether the output capability of the subsystem or end item must be increased.

From a personnel task performance viewpoint the time-line and associated analyses are important techniques for evaluating whether the available human aptitude and skill capabilities will be overloaded at any significant or critical point where human components are employed for accomplishing operational tasks.

For complex system designs it can be anticipated that it will not be unusual if the matrix should have as many as one hundred functional performance requirements parameters to be related to between ten to twenty subsystem or end item capability parameters.

The Apportionment of Reliability Values

The specific role that each system segment, subsystem, and component end item will play in providing the capability to perform its allocated functions must be described. The relative "load" that each functionally "active element" in the prime mission equipment–human operator operating functional sequence must sustain in carrying out a time-lined mission profile will determine the apportionment of the reliability value to be achieved in the selection and design of the technique for accomplishing the required functional transformation. The apportionment of reliability values is made in accordance with the "product rule." For example, for a function accomplished by two sequentially active elements where the reliability goal is 0.96, each active element must achieve a reliability value of 0.98 (.098 × 0.98 = 0.96), unless one of the active elements attains a higher reliability value, in which case the other one can have a lower value, provided that the product of the two values equals 0.96.

If the system engineer must deal with preselected equipment or human components to perform the transformation, he will have to evaluate their known performance reliabilities in relation to the reliability values to be achieved as derived from the functional requirements analysis. If the prescribed equipment and human capabilities exceed required functional reliability values, no further action is needed. If they are deficient, he must consider and present alternative solutions. These then become the subject matter for trade-off study to evaluate the possible alternative methods for fulfilling the functional reliability requirements in lieu of the predetermined technical approaches.

The Allocation of Availability Requirements

To attain specified system availability (integration of system reliability and maintainability) requirements, the contribution that each system element, subsystem, and component end item will be required to make to attain required system operational readiness conditions must be described. This will subsequently determine the detailed reliability, maintainability, and supportability design characteristics to be achieved in each system end item. It will also establish the detailed requirements for logistics support in terms of provisioning tools, test equipment, spares and repair parts, and consumables.

Trade-off Studies

As was indicated above, initial functional analysis must be based upon at least a general, gross concept of the performance and design characteristics of the system elements, subsystems, and component end items which will be employed for meeting system use requirements. As functional requirements to meet operational demands are defined in more detail, trade-off studies among alternative design approaches to satisfy these requirements develop progressively more detailed design solutions. These may confirm the original selections of system elements, subsystems, and component end item design approaches, or seriously question their appropriateness or adequacy. As the design becomes more detailed, the degrees of freedom for considering alternative design approaches decrease and the choice of solutions becomes more restricted. It is through trade-off studies that consideration is given to design features and characteristics which will facilitate the operational use of a system.

In the conduct of trade-off studies to determine the best system design solutions to meet given functional requirements, there will inevitably be a number of constraints to be observed. Some will be stated in the basic system requirements, such as performance, cost and delivery schedule requirements, and limitations. Others are inherent in the system design process itself in order to make the best use of various system element, subsystem, and component end item input–transformation–output capabilities and design characteristics when integrated in various combinations to produce a coherent system design.

Impact of System Performance-Effectiveness Considerations on Design Studies

Although we have been emphasizing the identification and definition of functional requirements as the basis for determining system design requirements because they furnish the means for obtaining consideration of design characteristics oriented toward facilitating the effective use of the system and its end items, actually, operational use requirements constitute only a fraction of the total design requirements. The main source of design requirements of necessity are derived from the availability of the various physical, electrical, chemical, mechanical, physiological, psychological, and other properties embodied in the elements, subsystems, and component end items of a man–machine system which will determine its capability for accepting inputs and transforming them into desired output performance characteristics under specified environmental use conditions.

In considering possible man–machine system design solutions, use of available input–transformation–output devices may be constrained by requirements for incorporating specified operability, maintainability, safety, and reliability design characteristics in order to achieve specified system performance-effectiveness criteria or standards.

Operability.   The main concern under this heading is with the input–transformation–output interrelationships for coherently combining equipment and human components into performance-effective and efficiently functioning units. Especially important are the provisions for sensing information which reveal system operating conditions and transmitting it as clearly understood intelligence to the human components about equipment status as required for effective decision making and for easy control of system output performance. From the standpoint of effective utilization of human components in a system, this is the classical concern of *human engineering.* In complex systems it is essential that the input–transformation–output interrelationships between equipment and human components be precisely determined and the resultant integrated performance requirements be engineered to produce a continuous dynamic functional process with a view toward reducing the probability of human error.

Maintainability.   Under this heading the interest is in those characteristics and features of design and the provision of resources which contribute to the rapidity, economy, ease, and accuracy with which a system can be kept in or, in event of failure, can be restored to normal operating condition in the planned maintenance environment in a manner that permits attainment of desired availability goals. Chief concerns are with such things as accessibility, transportability, repairability, checkout, monitoring, fault isolation, fault prediction, calibration, adjustment, serviceability, storability, tools, test equipment, consumables, spares, and so on. The impact of designing a system for ease of maintenance upon the overall performance effectiveness is more subtle than designing for direct operator–equipment interfaces, but it is equally important. Some errors in maintenance can be hidden completely from the inspection and checkout procedures. Any resulting indirectly caused catastrophic failures are just as costly as those directly induced by operator errors. Achieving a design which facilitates error-free human maintenance task performance requires the application of strictly controlled human engineering considerations in order to incorporate appropriate maintainability design characteristics in the various system elements, subsystems, and component end items.

Safety. Because of the dramatic quality of a catastrophic failure, especially if there is personnel injury or loss of life, safety engineering as a special design discipline has become a convenient avenue for forcing recognition of the importance of accomplishing effective system engineering. After end item designs have been put into production and after test or production quantities of hardware have been delivered, one of the best arguments to employ in support of a proposed design modification is to show a need to correct some unsafe condition. The safety ploy can be a potent one for obtaining design features and characteristics to improve system performance effectiveness which could not be achieved otherwise.

Reliability. By now it should be apparent to the reader that the objectives of human engineering, maintainability, and safety are all the same—namely to serve as a forcing function to institute an integrated system design and engineering approach such as we have been discussing. In complex system design and development projects, the reliability movement historically has most often had to shoulder the burden for promoting effective system engineering. This is reflected in the definition of reliability which is usually stated as "the probability that a system will perform a required function under specified conditions, without failure, for a specified period of time."

Performance of detailed parts, materials, and processes as incorporated in equipment end items was the first concern of reliability engineers. As collections of equipment were assembled and integrated in order to make them function together as a unitary system, the reliability analysts soon discovered that the test and failure data collected on individual component parts of the equipment did not necessarily predict their performance output characteristics and useful lifespan when they were integrated with other component parts and used in a given operational environment. "System reliability" and "operational confidence" naturally emerged as new concepts. This was approaching the system engineering problem by "backing into it" as a result of insights gained from clinical analyses of operational failures. Consequently, reliability engineers have increased their understanding of why systems fail. They have broadened the reliability concept to be more comprehensive in terms of incorporating it along with capability and maintainability considerations to compose the broader concept of system performance effectiveness. As a consequence, the system orientation of contemporary reliability engineering is reflected in the concern with such design characteristics as simplicity, vulnerability to damage, redundancy of functionally active elements, procedural adequacy and accuracy, and proficiency in critical personnel knowledge and skills.

Cost-Effectiveness Studies

Before arriving at an agreed-upon system design, by choosing between alternative scientific and technological approaches for achieving specified performance requirements, these alternative approaches must in turn be traded-off in relation to the customer's desires, whether they are explicit or implicit, as expressed in terms of cost and schedule requirements or limitations. The general guideline in the cost area seems to be that if a given system design will be "adequate" to do the desired job, then it is to be preferred to one which will cost more but would be "optimum" or perhaps "overdesigned" to ensure operational success regardless of cost considerations.

Formal "value engineering" programs are employed to sensitize all system project management and system design and engineering personnel to consider cost as a real and necessary system design parameter. The system engineer must be prepared in the course of trade-off studies to discuss the relative costs of proposed alternative design approaches. The cost estimate includes consideration of all life-cycle costs, namely, initial design, production, and testing costs, as well as operation, maintenance, and logistics support costs for the time period which is considered to be the useful life of the system. From a system performance standpoint the absolute cost of any particular element should never be considered in isolation. The cost of a given design approach should always be evaluated in relation to its contribution to overall system performance effectiveness measured against the cost of the alternative design approaches in order to achieve a comparable, a greater or a lesser degree of overall system performance effectiveness. Ultimately, the customer must decide whether the attainment of a given system performance capability is worth what it will cost him.

Development Schedule

Establishing development schedules is good system project management practice from a strictly business standpoint. From a system design standpoint, the system engineers must contribute to the preparation of such schedules, in terms of furnishing estimates based upon technical knowledge and judgments concerning necessary sequencing of events and the times required to design, fabricate, assemble, integrate, and test a system preparatory to deploying it for operational use.

Figure 5 shows a typical system engineering activities flow diagram for system acquisition. The label bars at the top of the diagram orient the sequential activities to normal configuration management baselines. *Functional configuration identification*

Figure 5. (see foldout opposite) Flow diagram of typical major system engineering activities for system acquisition.

covers the period of requirements and design analyses based upon a system specification to derive and allocate performance and design requirements to end items. It is completed when end item development specifications are issued. *Allocated configuration identification* covers the period of end item design and development and is completed when engineering and qualification analyses and tests demonstrate that end item designs fulfill their respective development specification requirements. *Product configuration identification* normally overlaps allocated configuration identification since it begins with the initiation of product specifications against which operational test items are produced and tested. These specifications in preliminary form are usually reviewed as part of the end item critical design reviews.

During the functional configuration identification period, a *system design requirements review* is accomplished prior to completion of end item development specifications to evaluate and refine the system design approach in relation to established project cost-effectiveness objectives.

The *preliminary design review* is conducted as early as feasible after issuance of end item development specifications to systematically relate the proposed designs to performance and design requirements. Development test plans culminating in verification and qualification of end item designs in relation to specification requirements are also presented. Further, cost-effectiveness analyses projecting the utility of the end item designs in their intended system application and operational use conditions are an important aspect of the design review.

The *critical design review* is held when engineering and qualification analyses and tests on end items are sufficiently complete to verify that designs meet specification requirements. Preliminary product specifications are reviewed at this point. Satisfactory completion of a critical design review results in release of drawings and detailed component design specifications for fabrication of first operational test, or production prototype end items. Acceptance testing of the end items is accomplished in accordance with production specification requirements.

The mainstream system engineering activities as shown in the diagram flow from the system specification through requirements analyses; system design optimization; end item development specifications; design reviews; and, test and evaluation for design verification. The iterative nature of system and development is shown by appropriate feedback flowlines.

Milestone schedules are prepared in relation to the acquisition flow diagram to show the actual time phasing of the activities for specific end item designs. End items employed in support of the prime mission end item tests normally will be produced earliest in a system acquisition schedule. Computer program designs, which are dependent upon test and control requirements built into the prime mission and associated support equipment end items, normally will be the last end items to be produced.

Important system engineering and design responsibilities to be accomplished in implementation of a system development schedule are discussed below.

System Design Description

The end product of the system design process is to synthesize a coherent total system design which will yield a defined set of operational capabilities from given inputs with respect to stated performance, cost, and delivery schedule requirements. However, it is not just the physical design of the separate end items taken singly, or even collectively, which yields desired system performance capabilities, but it is their design features and characteristics which will enable them to *work together to attain a desired system performance capability as a functional whole.* It is this designed-in *wholeness* which makes the system approach unique compared with designing an individual item of hardware by applying some given scientific and technological principles, regardless of how simple or complex its resulting operating characteristics may be. As an individual item of hardware, it accepts specific input values and performs a transformation role to produce specific output values without regard to its utility in a system application. As we have shown earlier in this discussion, it is the ordering of the individual end item input–transformation–output capabilities into a chain of functional events, including the use of human components both in series and in parallel, that results in a *system performance.*

Although it is a system engineering goal completely to define and describe the total performance and design requirements for all system elements, subsystems, and component end items prior to undertaking their detailed design, there may be situations where alternative design approaches may be tried and a decision as to which is to be finally selected deferred until after engineering test and evaluation. Also, in some systems, it may be necessary to compromise design objectives or performance requirements in order to attain some degree of early operational capability at the customer's insistence. On the other hand, acquisition may be delayed because key components may require development and testing to ensure successful application in the overall system. When such decisions are being considered, it is not unusual for some essential system design considerations from an operational applications viewpoint to be given short shrift. With the system engineering approach described here, and the resultant documentation of technical decisions as the basis for arriving at them, it is at least possible for a system engineer participating in a system design analysis effort to record any dissents from proposed design approaches and hopefully have them considered in the normal course of events in the system design process.

System Definition

The system engineering and design decisions must be documented for procurement purposes. The instrument used is the specification. Whether we are concerned with government or industry procurement of systems, or system elements, it is desirable to employ a uniform specification outline with defined and standardized content by paragraph numbers. As a basis for delineating the specifications to be written, a specification tree is prepared to cover a complete system. Such a tree accomplishes two broad purposes. First, it identifies the work breakdown structure for accomplishing the design of the respective levels of system elements, subsystems, and component end items. Second, it permits procurement of the system by convenient logical segments or work packages. The overall system specification, however, should be complete, comprehensive, and sufficiently detailed to define all of the performance and design requirements for all essential elements or subsystems of a system, thus describing their complete integration into a system which can be employed as an *entity* to accomplish the defined mission. Writing such a specification is an art in itself. If sections of it are written separately, it can merely set forth a collection of discrete subsystem performance requirements. To be a forceful and realistic system specification, each and every requirement must be evaluated in terms of its sensitivity to variations in the other requirements with which it is associated. It does not make any sense from the viewpoint of the systems approach to specify the attainment of absolute values in two or more system performance parameters whose interactions produce a combined effect upon the attainment of desired overall system performance effectiveness. The performance requirements should be stated in terms of the interdependency of such values to achieve a combined system output value.

End Item Design and Development

End item design and development is initiated when a system design approach is selected and end item performance/design requirements specifications are prepared. System design and engineering activities are concerned with monitoring the implementation of end item specifications. Any proposed changes to system and end item design requirements must be evaluated, no matter how minor the proposed changes may appear to be, to ascertain their impact upon the total system design. Otherwise, the integrity of the system design cannot be preserved.

**Test and Evaluation of Integrated System
Performance Capabilities**

One final task remains for system engineering to accomplish, and that is to specify proper instrumentation and test procedures by which appropriate data can be acquired for evaluating observed system performance in relation to the fulfillment of specified performance and design requirements. For a complex system composed of numerous end items, all of which must perform their assigned roles by interfacing correctly to yield the desired system performance capability, designing an integrated system test program is a major system engineering effort. The requirements of such a program are discussed in Chapter 6.

6

Testing and Evaluating
System Performance

What Is "Proof" of System Performance?

The nature of proof in the philosophy of science is a subject about which there have always been highly involved arguments. It is equally difficult to obtain universal agreement for a list of criteria which can be employed as a basis for selecting or designing tests for the purpose of proving that a complex system meets the specified performance requirements. There are two classes of tests with which we must be concerned from a system design and engineering viewpoint. One type of testing is aimed at *predicting* performance, and the other type is for the purpose of *verifying* performance.

Another basic concept which is important for designing a system-oriented test program, especially for large-scale complex man–machine systems, is that the dynamics which produce given performance characteristics are determined by a combination of *deterministic* and *probabilistic* input–transformation–output interactions of its functional components. It is essential, therefore, in designing a series of tests for acquiring data on system performance capabilities, that a careful identification be made of the type of measurements needed to supply the correct values for the particular mathematical and statistical models to be employed for analyses of the data. Making such determinations at an early stage in a system design makes it possible to provide appropriate test data pick-off points in equipment components, either for feeding built-in test devices or for connecting external test equipment.

Increasingly, as a consequence of the capability provided by microelectronic circuitry, built-in test and control devices become feasible for incorporation in system designs. When such capabilities are applied properly, they can enhance system performance dependability by permitting detection of early degradation of component performance for instituting corrective action prior to its failure. However, we are not concerned here with the details of implementing system designs to facilitate system testing and control. Our interest is in identifying and describing the tasks to be accomplished by system designers and engineers in the test and evaluation area. The fundamental questions are, "What is the purpose of the test, and what will it prove in relation to predicting or verifying system performance?" and "Do we need this test, or what risk will we incur if we do not perform it?"

Integrated Test Program Objectives

The test program should provide incremental integrated testing so that each test constitutes a step in providing data which collectively increases confidence in predicting successful mission accomplishment. This approach requires a carefully designed data recording, compiling, processing, and retrieval system which can be readily updated and interrogated at will for detecting:

1. Designs and design changes which fail to satisfy stated performance capability requirements or design characteristics for operational use;
2. Fabrication or assembly errors; and,
3. Component failure or degradation which prevents or compromises proper system performance.

The collection and correlation of data should start with initial parts, materials, and processes testing and component tests, and then continue systematically through integrated system tests and operational readiness demonstration. When carefully planned, an integrated test program can ensure complete testing without unnecessary duplication of test data. The program culminates in complete operational system tests under environmental use conditions which exercise all prime mission equipment and associated support equipment, procedures, personnel skills, and logistics support. Such testing is conducted over the complete spectrum of normal and abnormal operating modes. It should verify that no combination of expected demands upon system performance response capability causes a system failure.

Here is the progression of how an integrated program for testing proceeds:

Parts, Materials, and Processes Testing. When parts, materials, and processes with known histories of successful application in systems similar to the one being designed are to be employed, emphasis in testing can be placed upon ensuring that they will continue to demonstrate the same performance and design characteristics that have been demonstrated previously. When new parts, materials, or processes are required because of unique system requirements, adequate testing should be accomplished under environmental use conditions to demonstrate that these new factors can meet all specified performance and design requirements as appropriate for a given system design.

Component Testing. From a systems viewpoint, laboratory constructed "breadboards" or "brassboards" of proposed electrical and mechanical components should be designed, built, and tested only when the needed technological knowledge to be gained cannot be obtained in any other way. Generally, breadboards or brassboards cannot be reproduced in the normal factory manufacturing process and, therefore, the test data obtained from breadboard or brassboard testing will have limited value for predicting component performance as it will be configured for fabrication and incorporation in the next higher assembly and used in the operational system environment. Most design effort at the component level can be adequately supported by mathematical analyses and computer programs which simulate input–transformation–output values, rather than by empirically arriving at adjustments of such parameters by applied experimentation. Producing an actual prototype model of a proposed component and then testing it under simulated environmental use conditions employing a full dynamic range of variable input and output parameters yields generalizable data at a high level of confidence for predicting probable performance in its intended system application. Furthermore, as in parts testing, prototype components can if necessary be subjected to testing under adverse environmental or extreme performance limit conditions to establish what "overload" or "overstress" conditions they are likely to withstand for reliability rating purposes. Performance anomalies which occur under test conditions should be thoroughly analyzed for possible impact on system performance. Design changes and contingency operating procedures can be generated as a result of such analyses. Off-the-shelf components which have been used successfully in system applications with similar characteristics as the proposed new design do not normally require retesting. If they are to be subjected to new demands or environmental circumstances, then it may be desirable to test their performance responses in relation to the specific conditions which will be encountered in the new system.

End Item Testing. End item testing should be restricted to:
1. Demonstrating the physical and functional integration of the various components as assembled into the end item to produce specified input–transformation–output functions under simulated operational load conditions.
2. Conducting tests of any special environmental effects such shock and vibration, thermal ranges and changes, radiation, corrosive substances, moisture, and humidity, which simulate predicted operational use conditions.

System Testing. Normally, subsystem testing is neither realistic nor cost effective. By definition, a subsystem is only a functional entity rather than a self-contained physical package (such as is a component end item). The several separate components which through their interactions produce a subsystem performance capability are usually configured with some components which are employed in other subsystems. Accordingly, rather than simulate the other subsystem actions and their environmental effects for testing purposes, it is best to assemble and perform system integration and testing on all end items functioning in their intended use environment, including use of the human components employing normal operating procedures. Such testing will prove out both subsystem and integrated system performance capabilities under realistic combined internal stress and environmental effects conditions, including simulation of both normal and abnormal variations of such conditions. Provided that the sequence of tests and analyses have been appropriately selected, the cumulative data should provide evidence that specified system input–transformation–ouput capabilities have been attained. Prediction of probable system mission success can be made at least with some degree of confidence.

Operational Data. With a data base established in the integrated test program for demonstrating performance capabilities and limits, additional data can be entered as it accumulates from operational use of the system. On an "as needed" or a periodic basis, such data can be summarized for evaluation. Strong and weak points in the system and equipment designs can be identified for use in new development projects for similar systems or for proposing modifications in existing designs when it would be cost effective to do so.

Data Analysis and Evaluation

If there is to be a system-oriented integrated test program, then it must follow that the collection, analysis, and evaluation of data from such a program must be planned so as best to predict and subsequently to verify system performance goals as design and development progresses. Table 2 presents in a system-oriented manner the essential building blocks for such a data collection, analysis, and evaluation program. Column A of Table 2 divides the table into the normal system project management activity phases. Column B lists the generally accepted system design and development steps from project inception to system operational development. Currently, the only argument about the sequence shown is the extent to which the several consecutive, sequential steps can be overlapped or accomplished in parallel by substituting engineering analyses for testing and, therefore, develop, procure, and deploy an operational system on a "concurrent" basis. Actual experience under the "concur-

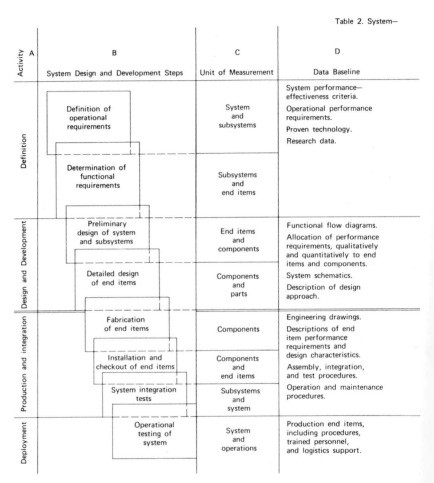

Table 2. System—

Activity	System Design and Development Steps (B)	Unit of Measurement (C)	Data Baseline (D)
Definition	Definition of operational requirements	System and subsystems	System performance—effectiveness criteria. Operational performance requirements. Proven technology. Research data.
	Determination of functional requirements	Subsystems and end items	
Design and Development	Preliminary design of system and subsystems	End items and components	Functional flow diagrams. Allocation of performance requirements, qualitatively and quantitatively to end items and components.
	Detailed design of end items	Components and parts	System schematics. Description of design approach.
Production and integration	Fabrication of end items	Components	Engineering drawings. Descriptions of end item performance requirements and design characteristics.
	Installation and checkout of end items	Components and end items	Assembly, integration, and test procedures.
	System integration tests	Subsystems and system	Operation and maintenance procedures.
Deployment	Operational testing of system	System and operations	Production end items, including procedures, trained personnel, and logistics support.

rency" concept has demonstrated beyond a reasonable doubt that none of the steps as shown can be omitted, skipped over lightly, or completely telescoped. When thoroughness in accomplishing the work in any of the steps has been neglected, it has always been necessary at some point in time to "backtrack" and accomplish it. The usual and all too familiar symptoms of failure to follow a rigorous scientific approach in system design and de-

Oriented Data Analysis and Evaluation

E Methodological Approach	F Quantitative Values Employed	G Type of Statistical Verification	H Measurement Objectives
System performance—effectiveness mathematical models. Functional requirements analyses. System schematics showing subsystem and end item functional interactions. Computer simulations of input—transformation—output interrelationships. Design and trade—off analysis.	Figures of merit employed as criteria for evaluation of system effectiveness of alternative system design approaches.	Predictions of the relative probabilities of achieving desired performance capabilities by selecting various combinations of discrete design approaches with assumed output values and correlated interactions.	Quantification of Input–Transform–Output Units for Employment as System Building Blocks
Engineering drawings. Static (soft) mockups. Testing of development models by dynamic simulation of functional performance under operational load and environmental use conditions.	Standardized numerical values on calibrated scales used to measure output responses of designated units to controlled input stimuli.	Replication of measurements until preselected confidence criteria are satisfied in order to establish that obtained values fall within prescribed limits.	
Exercising end items, both singly and as integrated in their intended system configuration, under defined environmental conditions and operational use loads.	Record of events on a real time baseline and quantitative descriptions of achieved performance outcomes.	Attainment of specified terminal quantitative performance outcome under defined environmental conditions, applying the null hypothesis for deciding the level of confidence to be ascribed to the values obtained in relation to the number of successful replications achieved.	Quantification of Functional Events in the Sequence of System Operations
Status monitoring to verify performance effectiveness of system concept and design.	Distribution of times between failures of like type.	Probability of occurrence of failures by type.	

velopment are degraded system performance, delivery schedule slippages, and excessive costs.

System design and development, to be ultimately effective, must be oriented toward producing an operating end product. Column D of Table 2 shows that, in relation to the development steps, the path to demonstrating the fully operating system end product starts with established firm system performance-effectiveness criteria and operational performance requirements, both qualitative and quantitative. These are employed as the standards for evaluating the system end product performance measurements obtained in the integration and operational test phases. From initial definition of system requirements to complete operational testing of a system involves a progressively more detailed technical data baseline, as shown in Column D. As shown in Column C in relation to the successive design and development steps in Column B, the unit of measurement likewise becomes progressively more definitive, detailed, and precise through the completion of detailed design of end items, which is shown as the double "midline" in Table 2. The data baseline (Column D), against which evaluations of the products of each of the requirements and of the design and development steps (Column B) are made, reflects this increasingly exact determination of input–transformation–output performance values to be achieved in the selection of end items, their components, and parts, materials, and processes in order to achieve progressively system, subsystem, end item, and component designs. Before proceeding to discuss the material presented in Table 2 below the "midline" separating the phases concerned with system definition, design, and development, from those for system production and testing, let us proceed horizontally across the chart.

Column E in Table 2 lists the primary or major methodological approaches which are employed for making system engineering decisions in relation to the system design and development steps (Column B) where they are most applicable. For example, for formalizing system performance-effectiveness criteria, the primary methodological tool is a system performance-effectiveness mathematical model. As a system engineering technique, employing a mathematical model permits studying how variations in characteristics affecting performance capability, reliability, and maintainability will most probably influence the attainment of desired overall system performance effectiveness.

From the very inception of system planning, a system performance-effectiveness mathematical model is oriented toward the user's requirements for deploying a system to accomplish a specified mission. Selecting realistic values to be employed as figures of merit in a system performance-effectiveness model is essential. To do so requires using the definitions of the system functional requirements. For example, if the prime mission equipment is to be used only for a very short period of time, and is, therefore, for all practical purposes rapidly expended, then it does not make sense to assign a high value to the maintainability factor in the system performance-effectiveness model. Conversely, if it is to be reused repeatedly to perform its stated mission, then a high value should be assigned to the maintainability factor, as well as to reliability.

By progressively allocating quantitative system performance-effectiveness figures of merit to ever more definitive "functional blocks," which represent the employment of the equipment and its human components in a coherent system design, and then by making trial design solutions, the resultant alternative combinations can be processed by computers to analyze the effect of varying the input–transformation–output values for various equipment component combinations upon the overall system performance effectiveness in relation to achieving specified mission outcomes.

A computer simulation capability for studying system performance effectiveness is important from a system engineering standpoint for a number of reasons.

1. With the increasing complexity and cost of large-scale systems, plus the importance of producing desired performance capabilities on preestablished development schedules, we are being forced increasingly into collecting and analyzing test data on a "bits-and-pieces" basis. In any system design and development project, only a few very complete systems can be made available as strictly developmental test items. A fully structured system performance-effectiveness mathematical model can be employed for progressively evaluating design approaches in terms of predicting performance with increasing statistical confidence as the "bits-and-pieces" of test data become available. The types of statistical verification which must ultimately be achieved in relation to each major developmental activity are shown in Column G of Table 2. All of the design and development test data collected on a piecemeal basis must finally fit together like the proverbial

jig-saw puzzle at the end of the integrated system tests to reveal whether or not it is advisable to proceed with system development with reasonable confidence in achieving successful mission outcomes (including any planned, graceful, and beneficial degradation in order to sustain a partial capability in the event of irreparable component failures while performing an operational mission).

2. Employing computer simulation programs in conjunction with testing of end items under dynamic real-time simulation of combined operational loads and environmental use conditions during the system design and development steps, as shown in Column E in Table 2, to study the application of various combinations of subystem end items and components to carry out simulated mission operations, permits implementing an integrated system design approach for deriving initial system designs which cannot be achieved by waiting until all end items are individually designed and produced and then attempting to integrate and test them. High costs, systems complexity, and the urgency of developing new systems quickly, are actually working in favor of winning acceptance and implementation of system engineering methods and procedures aimed at achieving integrated total system designs. Through increased use of simulation programs, alternative design approaches can be studied in depth prior to committing a given approach to detailed design and fabrication. For example, both equipment- and personnel-oriented system engineers are able to take a common look on a functionally equivalent basis at a given man–machine system design approach. They can work with the same trade-off considerations in studying alternative design approaches for combining equipment and human components, using the same common system performance-effectiveness mathematical model to evaluate their probable impacts on system outcomes.

3. Finally, computer simulation gives us the capability of studying where component failures, whether human or equipment, will be critical for degrading or aborting a system mission. This knowledge then allows concentration on achieving optimum designs to meet functionally critical requirements, and to plan the most effective programming of equipment and system tests

particularly in terms of obtaining the most significant data for evaluating the adequacy of system design solutions in "high risk" areas. With the kind of visibility that a planned progressive evaluation of system performance data permits, the use of a very small number of actual test items, produced and tested in their intended operational configuration, results in placing increased statistical confidence in a very few replications of successful performance demonstrations.

Column F in Table 2 describes the quantitative values that can be employed in relation to evaluating the data which become available during the respective system design and development phases. In the earliest phases of conceptual designs, figures of merit can be established as a basis for evaluating the probable effectiveness of alternative design approaches. This requires the assignment of quantitative values as an estimate of the probable performance characteristics of the various design approaches which are to be manipulated. Some of these values can be based upon considerable previous experience in applying a given technology to produce a particular type of output under specified conditions. Other values will be merely best estimates with little confidence initially in their validity. However, it is necessary always to attempt to make all estimates on as reasonable a basis as possible in order to try to prevent significant overestimating and underestimating of predicted performance capabilities.

When estimates of performance characteristics are related to probable cost and delivery schedule estimates for the various design approaches being studied, it is also possible to estimate the relative cost effectiveness of proposed alternative design approaches. It is emphasized that initial results from conducting mathematical model exercises for estimating both system performance- and cost-effectiveness early in a system design will yield only gross-level predictions. However, by following through with the compilation of increasingly detailed and precise data as system design and testing progresses, the numerical values become more accurate. With replication of stable performance measurements obtained in a system test program, the statistical confidence in their dependability increases.

As indicated in Column H of Table 2, the measurement objectives during the design portion of a system design and development project are to establish quantitative values for the various functional elements of a system for describing their input–transformation–output performance characteristics as building blocks in a system design in relation to predicting their impact upon system performance. Below the midline in Table 2, the interest shifts to the quantification of functional *events* in the sequence of system operations.

After components and end items are fabricated, assembled, and integrated for system testing, the testing interests centers on demonstrating that their collective employment produces the specified system performance capabilties. The first objective of system testing is to detect any *systematic* causes of failure which require design changes to eliminate.

Failures or anomalies in system performance due to *random* causes require clinical investigation and analysis to determine their cause. Large-scale complex systems have innumerable sources of malfunctions which can cause unwanted performance anomalies or failures. The chances for their occurrence, just as are the chances for obtaining correct performance, are a matter of statistical probability. Whether or not a cause of a given random failure should result in a design modification is a system engineering problem which involves all of the system performance- and cost-effectiveness considerations that are employed in the decision-making process for arriving at initial designs. In fact, a decision to effect a redesign to remove the cause of a given failure may have a "chain-reaction" effect which will introduce more chances of failure occurring in other portions of the system than if the risk is taken of a given failure recurring in the normal course of events of system operations.

Throughout the integrated test program for system design and development, the system performance-effectiveness model is employed as the basis for planning the collection, analysis, and evaluation of test data. Without such a model, measurements of performance cannot be interpreted in terms of their value for predicting or verifying the attainment of desired system performance characteristics.

Ultimately, as shown in Table 2, a system test program culminates in demonstrating system performance capabilities and successful mission accomplishment under operational environment conditions. The measurement objective is to accumulate successful attainments of specified system performance and mission goals, as defined in the system performance-effectiveness model. Confidence in failure-free system performance increases as the number of successful demonstrations of system performance capabilities increases.

Collection of system performance data continues during system operational use. Generally, during system testing prior to its deployment, it is not possible to achieve fully all possible variations in performance demands as may be encountered in committing a system to accomplish its intended mission. Consequently, as shown in the bottom line of Table 2, interest continues for collecting data which verifies the performance effectiveness of the overall system concept and design. As in formally conducted operational system testing under controlled conditions, attention is centered on analyzing causes of any failures. If certain failures of like type occur with increasing frequency under given operating conditions, it must be determined whether or not there is a design weakness which requires a corrective modification in all deployed systems on a retrofit basis. Collection and analysis of system performance data normally continue until a given system is considered obsolete and, therefore, until there is no value to be gained from changing its design characteristics.

Analyzing and Evaluating the Contribution of the Human Components to System Performance

There is a very simple clear-cut criterion for deciding when human task performance is functionally equivalent with equipment task performance in carrying out the sequence of events which culminates in accomplishment of a system mission. It is as follows: Human tasks are functionally equivalent to equipment performed tasks when there is no subsequent step in the sequence of events that occur prior to mission completion where a checkout or test of the prime mission equipment status inherently verifies accurate human task performance. This is so because the correct performance of such human tasks places or keeps the prime mission equipment in its normal operating status and, consequently, they are functionally critical because their accurate performance is essential for successful mission accomplishment. Conversely, failure to correctly perform the tasks will result in mission failure or degradation. Conceptually, this is the same criterion as is employed by strictly equipment-oriented reliability engineers to identify the "active elements" in an equipment design upon which the functional reliability of its performance depends.

Equipment components or end items can be designed to perform input–transformation–output functions within very narrow tolerance limits. Human components on the other hand have a much larger physiological and psychological functional capacity than normally is required to perform their assigned tasks in a typical system operation. It is this reserve capacity and the self-adaptive capability possessed by humans that make them ideal components to perform certain tasks if the appropriate conditions to facilitate their correct accomplishment are properly designed into a system.

There are a number of considerations which determine the best use of human and equipment components in system engineering trade-off studies. A system design team which is concerned with predicting potential sources of system error must consider the possibilities for human error when weighing the use of human components against the use of equipment components in alternative design approaches. Such a team will be composed of technical specialists who collectively will be thoroughly knowledgeable concerning the advantages and disadvantages of using human or equipment components in terms of their relative error-producing potentials for performing system tasks under various environmental use conditions.

In constructing a system performance-effectiveness model and in allocating figures of merit to express the relative values of various system components for accomplishing specified mission functions, it must be borne in mind that personnel performance is always a "load bearing" factor to some extent in achieving a quantitative system performance output. As has been discussed above, human performance contributes directly, without equivocation and as a statistically independent factor, to any quantitative measure of system performance effectiveness when dependence is placed upon personnel task performance as an active element in a sequence of functions required to accomplish a mission objective. In the system design process the human factors-oriented system engineer deals with the same kind of statistical probability problems in predicting human task performance that hardware system engineering specialists face in predicting equipment performance reliabilities. The technology needed to accurately apply the use of human components in system design is as well understood as is the application of mechanical or electrical components. Perhaps we are confronted with a higher level of uncertainty in making predictions for given system applications because of the more complex and variable nature of the performance of the human components that are to be employed. Unless well-trained and self-disciplined in their task performances they can constitute a highly variable and undependable quantitative performance factor. Statistically, therefore, the standard deviation of the measurements of human input–transformation–output functions can appear to show gross variations in performance characteristics against the fine variations which are employed to describe allowable tolerance limits for performance of equipment input–transformation–output functions. However, this does not mean that the available technology upon which to base decisions for employment of human components is any less exact than that upon which decisions are made for employment of mechanical and electrical components. The allowable tolerance limits generally need to be larger in the design of man–machine interactions than in functional interfaces between equipment interfaces. The system engineering role is to state exactly in input–transformation–output terms how the various equipment–human component functional interfaces must be designed so that any predictable human variances in performance output has as low as possible probabilty for causing a system performance failure. Attempting to eliminate reliance upon correct human task performance in designing given system operations is not always realistic. For example, replacing a requirement for relatively simple straight-forward human performance of certain input–transformation–output functions with equipment designed to accomplish the same function may require a complex, relatively unreliable equipment design. In such a case the trade-off considerations in relation to system performance- and cost-effectiveness criteria will obviously favor the use of the human component.

Analyzing and Evaluating the Contribution of Support Components to System Performance

In a total system design, standardized operating, repair and testing procedures and support equipment end items are designed to be employed by the human operator, controller, or maintainer of the prime mission equipment. Likewise, facilities and facilities equipment are designed to provide appropriate housing and controlled environmental conditions, if required, for the proper conduct of operation, control, or maintenance functions of the integrated man–machine system. Most failures in the procedures, support equipment, and facilities components can be detected and handled independently of controlling the performance of the prime mission equipment. However, some failures of the latter can be caused by failures in the former components to provide the proper interface conditions which are required to sustain the normal operating condition of the prime mission equipment. Provisions must be made in a system test and evaluation program to obtain and analyze data to predict and verify the proper system performance of procedures, support equipment and facilities components.

Implementing System Test and Evaluation for Design Verification

It is a significant system engineering responsibility to plan, monitor, and evaluate the results of an inegrated system test program to verify that specified performance and design requirements have been met. Table 2 presents the methodological basis for such a program, Figure 6 diagrams the time-phased activities which are normally involved in a system test and evaluation program. As shown on the diagram the major functions are: test requirements; test planning; test readiness; test operations; test review, analysis, and reporting; and, test evaluation. Each of these functions are described below.

Test Requirements Analysis. The systems approach to test and evaluation consists of a series of steps in determining the capabilities of equipment and any associated software to perform within the prescribed design envelope, and to evaluate potential performance in relation to specified requirements. In so doing, a test requirements analysis, including performance- and cost-effectiveness considerations, is accomplished. The test requirements analysis begins with a systematic evaluation of system and end item performance and design requirements as set forth in the system and end item specifications; the available test techniques suitable for application to verify the requirements; and, the spectrum of constraints which must be observed in implementing the test program. The analysis results in a concept for testing and evaluating end item performance and the pyramiding of end item tests into system integration and operational tests. In addition to system and end item specifications, other test requirements covering personnel, support, and procedural requirements are derived from appropriate project development plans covering such subjects. Special test requirements and procedures are derived from system-effectiveness requirements documents covering such subjects as: availability/reliability/ maintainability; human engineering; safety; value engineering; and the like. Configuration management procedures are important from the standpoint of maintaining proper control of any specification changes resulting from the test program. Quality assurance maintains surveillance over the test program and verifies that all test requirements and procedures are properly implemented.

Test Planning. The capabilities of a system can be demonstrated to some degree at every stage of design and development. With

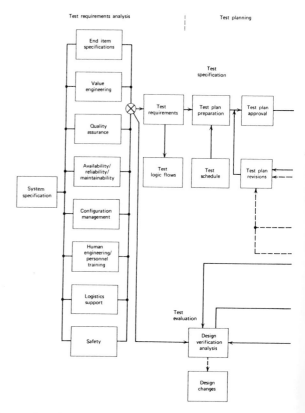

Figure 6. Test and evaluation activities flow diagram.

careful and early planning, a spectrum of tests can be identified that will assure verification of certain elements of design very early in the development stage; thus providing a strong data base upon which to design (or redesign, if necessary) interacting system elements. Examination of the system components piece by piece will allow the identification of those elements that lend themselves to early testing. It will also identify those physical elements or operational functions that are more complex because of their required interdependence to produce a specified system performance capability and, therefore, would more logically be tested later in the development program. Following this approach

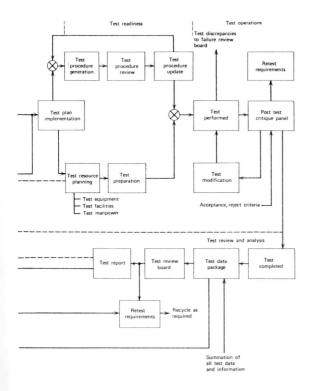

Figure 6. *Continued*

will provide a test data base that accumulates as each test is completed until total system performance is verified.

Test Readiness. The implementation of the actual testing begins with the establishment of test readiness when all test procedures and resources have been assembled.

Test Operations. Conduct of tests will be in accordance with the published procedures. A test conductor will be responsible for seeing that each test is carried out properly. At the conclusion of each test, the results of each test are reviewed by a post-test critique panel. This panel will be responsible for determining that the test was correctly performed. Test anomalies should be identified, and, if necessary, the panel should direct a retest or a test modification.

Test Review, Analysis, and Reporting. During the test operations, test data will be packaged for review and analysis by system engineering and by cognizant functional engineering groups representing all applicable technologies and design disciplines. Any performance anomalies or design deficiencies revealed by the test data will be studied and analyzed. Test reports incorporating the results of the analyses will be prepared and will recommend any design modifications deemed essential for specification compliance.

Test Evaluation. Test data, analyses and reports will be further evaluated to assess the performance of each end item, both individually and collectively, as integrated into an operational system. This step completes "closing-the-loop" in the design verification process. Design verification as a continuous iterative process will have begun with early analyses of proposed design approaches, and will have proceeded through engineering analyses; breadboard/brassboard testing when applicable; engineering tests of components and end items when necessary; qualification tests of end items; system integration tests; and, finally, operational testing of the deployed system. The evaluation will normally arrive at conclusions covering such things as:

1. Proof of designs for fulfilling specified operational and mission requirements in a performance-effective manner.
2. Evaluation of personnel and logistics support requirements as a basis for assessing cost of ownership over the expected useful life of the system.
3. Recommendations for design changes to improve end item and overall system designs.

The Proof of a System Is in Its Use

What does it all prove? Conducting an integrated program of system testing and evaluation to predict and verify its performance capabilities is essential. However, the only real proof of a system's utility comes from the customer's satisfaction with its quantitative and qualitative performance and design characteristics. If he is satisfied with it in terms of economically and conveniently accomplishing his use objectives, then it is a good system. On the other hand, if it fails to achieve his performance goals or is deficient and perhaps costly in some aspect of operability, reliability, maintainability, or safety design characteristics, then from his viewpoint it can be a bad system.

A brief quotation of a terse comment by Admiral H. P. Smith of the United States Navy illustrates this point as follows:

> My ships are burdened with so-called sophisticated equipment which have wonderful "press clippings" concerning their performance. Unfortunately, they won't work when we need them. Those complex systems are generally unreliable and very difficult to maintain. When they work their performance is usually quite good. However, I would gladly sacrifice some performance for the sake of reliability and maintainability. My ships need systems that work when they are needed to work. They don't need any more junk installed in them.[*]

[*]Smith, Harry P., quoted in Jayne, Gordon H., "Planning Integration in System Design," Northeastern States Navy Research and Development Clinic, *Proceedings*. Philadelphia, Pa., November 1964.

7

Computer and Software System Engineering

Characteristics of the Computer/Software Development Process

Nearly all complex systems involve one or more computers for processing information for operational control purposes. Also, a computer can be employed by itself as a purely information processing system. In either case, the role of a computer in a system is dependent upon a programmed set of instructions which tells it how to process signals fed to it and how to output the processed results. The design, development, and checkout of a computer and its software, therefore, constitutes an important, and more often than not, the key element in the successful design of a man–machine system.

Designing a computer controlled system involves devising ways to execute any number of input–transformation–output functions, either in series or in parallel. Designing a computer program to manage such functions requires the application of pure logic for devising the information processing routines. Specific procedures, called *algorithms*, are employed to define the problem-solving transformation functions. They establish how given inputs will be handled by the computer to produce the desired output information. Because computer programming is strictly an "intellectual" exercise, its effective system engineering is the most difficult aspect of designing a complex man–machine system.

The design, development and checkout of computer programs usually proceeds for a considerable period of time after the computer hardware designs with which it is associated have been operationally verified. This occurs because computer program testing can never fully exercise all possible "logic paths." Operational use will employ the untested logic paths and errors can be normally expected to be revealed in them. Also, new operational applications of a computer-driven system will require modifications of its software. Consequently, development of computer programs for a given system will never be completed as long as variations in its operational applications can be conceived.

Basic Functional Description of a Computer

General purpose computers differ in size, form, and methods of operation. However, they all possess certain common features. These are diagrammed in Figure 7. As shown in the diagram, a computer must have an *input device* for insertion of information and instructions. There must be a device where the processing operations as directed by the instructions can be performed on the information entered. This is the *computing unit*. The computer must have a capability to remember information and instructions, hence, a *memory unit*. Then, there must be a *control unit* to manage the execution of the step-by-step procedures for manipulating the input information in accordance with stored processing

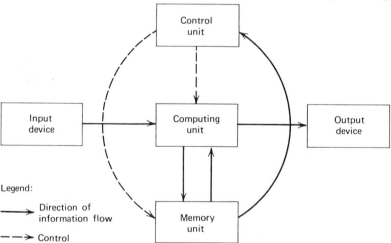

Figure 7.   Basic functional design of a computer.

instructions. Finally, data processing results must be delivered through an *output device.*

Admittedly, this is an oversimplified description of a computer. However, Figure 7 does describe the basic functional design for all computers, whether all functions are incorporated in one "box," or whether many cabinets of specialized circuitry are needed for various input and output devices, multiple data storage units, and specialized system controls. With the basic concept of a computer in mind, we can define more explicitly the various elements of a computer/software system.

Definitions of Computer/Software Elements

Before proceeding to discuss computer/software system engineering considerations, it will be helpful to define some basic elements of information processing systems.

Operator Console Input Devices. These are employed to introduce instructions and information required to initiate a processing function. Such devices include: *keyboards; switches; pushbuttons;* and *light pen.*

Input and/or Output Devices. Some devices can be employed either to make information inputs or to record processed information. These include: *punched cards; perforated tape; magnetic tape;* and, *magnetic disc* or *drum.*

Output Devices for Information Displays. These are employed to present the results of information processing for human consumption. Such devices include: a *printout;* a *data plot;* a *cathode ray tube; electroluminescent displays;* and, *transilluminated indicators.*

Central Processing Unit (CPU). The CPU or main frame is that portion of the computer which performs the calculations and decision functions.

Data Storage. The *memory* is that portion of the computer where information and processing instructions are stored. The *core memory* is the main memory and is normally the only memory directly accessible to the CPU. It is also referred to as being the *on-line* memory or *central data storage. Off-line* data storage or memory is employed to perform special purpose data handling functions. Central data storage can be programmed to be either *permanent* or *temporary.*

Data Base. The *data base* is that data stored in a computer memory in a logical or numerical sequence.

Storage Capacity. Storage capacity of a computer memory device is measured in *words* (also called *cells* or *registers*) which are usually of fixed length, normally consisting of twelve to forty-eight bits. This number is called the machine's *word length.* A *bit* (binary digit) is the minimum unit of information storage and has only two possible values, usually expressed as one or zero. Capacity can also be measured in *bytes,* which are units of six or eight bits, each employed to represent one alphabetic or numeric symbol.

Data Processing. *Data processing* involves the handling of information in accordance with precise rules and procedures to accomplish such operations as: classifying; sorting; correlating; integrating; comparing; calculating; summarizing; storing; and, recording.

Computer Program. A *computer program* defines the procedure which when reduced to a computer language enables a computer to accomplish desired information processing functions. A computer program is normally subdivided into *modules* composed of *routines* or *subroutines. Software* is the term used to refer to the totality of programs and associated operator instructions available for a computer.

Coding. *Coding* is the process of writing the detailed step-by-step instructions for the computer to follow.

Computer Language. The repertory of instructions available for translating a program into coded routines or subroutines is called a *computer language* in contrast to a *natural language* such as English. The set of instructions for a specific computer is that computer's *machine language. Higher-order languages* are designed to simplify the tedious aspects of writing a specific machine language.

Procedure-Oriented or Problem-Oriented Languages. Examples of such languages are FORTRAN, ALGOL, JOVIAL, and COBOL. The first three have been devised mainly for scientific data processing, and the last for business data processing. PL/1 is a language for both scientific and business data processing.

List-Processing Language. LISP is an example of a computer language devised for nonnumeric computations applicable to heuristic processes frequently employed in conducting research where precise methods of problem-solving cannot be predetermined but is discovered via a self-evaluative process as the program progresses.

Buffer. A *buffer* is a storage device used to compensate for a difference in the rate of flow of information or the time of occurence of events when transmitting information from one device to another.

Compiler. A *compiler* is a program which converts higher-order languages into a specific machine language. Programs which perform a similar function but at a much simpler level are called *assemblers.*

Parameters. *Parameters* are individual values which control the operation of programs or subroutine.

Some operationally oriented computer application terms will be defined in context as they are employed in the discussion to follow.

Determination of Computer Capacity

Information processing requirements have traditionally been employed for "sizing" computer capacity. However, with the advent of large-scale-integrated (LSI) circuit technology, and its resultant reduction in computer size, weight, power requirements, and cost along with increase in capacity, speed, and reliability, routine use of "oversize" computers becomes feasible. This permits flexibility and growth in computer application to enhance system operational capabilities via modifications of initial computer programs. This is important from a system engineering standpoint because initial sizing of information processing requirements invariably underestimates their ultimate magnitude once the operational utility of a system has been established. There is always an increase in the number of operations to be performed once the users become familiar with a system and are imaginative in conceiving extended applications for it. Consequently, overall system lifetime cost effectiveness will normally be enhanced by procuring a computer capacity, or by providing for growth in capacity, of at least fifty to one hundred percent more than that identified by initial sizing analysis. Modularization of computer designs permits increasing data storage and processing capacity by adding units to a basic computer configuration.

Investing in what may appear to be an excessive initial computer capacity in relation to estimated operational requirements does not preclude, of course, the need for sound software system engineering. One must always resist the Parkinsonian tendency to absorb an available excess capacity with marginally useful tasks. Computer programs cost money to develop and verify. Cost effectiveness in the choice of computer programs should always be a primary factor in determining whether or not to proceed with developing a given program. A man's time to perform the same task may be more cost effective and equally reliable in the long run. This is especially true when he must be present for the operation and needs to be constructively employed in performing it in order to enhance the motivational climate of his job by giving him something useful to do.

Basic Software Functions

Basically, system computer programming requirements are derived from operationally oriented information processing needs involving the use of computers for such things as:

1. Real-time information processing which provides a capability of immediate problem solving.
2. Static or "batch" information processing which involves assembling a number of input items for processing during a given computer run.
3. Keeping catalogs of information up-to-date.
4. Analyzing and correlating data from multiple input sources.
5. Comparing current information with stored reference information.
6. Presenting decision-oriented displays of information for problem solving.
7. Displaying operating equipment status information on a current basis for performance monitoring, diagnosis, and trend analysis purposes.
8. Controlling equipment operations or a dynamic process (e.g., cracking and distillation of petroleum products) either by manual input instructions or by automatic closed-loop signal analysis and command generation.

Additionally, the use of computers and computer programs to perform such operations requires control and support computer programs for internal control of computer operations such as the following:

1. Signal input processing.
2. Data output processing.
3. Computer input-output control.
4. Display control.
5. Processing control.
6. Data storage.
7. Programming verification.

It is not within the scope of this discussion to be exhaustive or to discuss in detail how to design computer programs. However, from a system engineering viewpoint it is essential to understand capabilities and the basic process of designing computer programs, and what are some of the commonly occurring design and development problems. Also, from a system standpoint it is impossible to separate software design from computer design. Both from a performance- and a cost-effectiveness viewpoint, their design must go hand-in-hand. This can best be illustrated by describing the kind of trade-offs which must be considered when selecting computer/software program design approaches. But first, the need for given computer programs must be established by requirements analyses.

Software Performance Requirements

Unless an information processing requirement is inherently defined by the characteristics of an equipment operation to be controlled, a software requirements analysis normally is initiated by preparing a *scenario* of the operation to be performed. In this scenario an attempt is made to present a nominal mode of operation to establish the baseline operational requirements. Subsequently, possible deviations from the nominal are studied and requirements are derived for contingency modes of operation to overcome, if possible, system performance anomalies (or, at least to operate the system in the most beneficial manner in a degraded mode). If corrective maintenance is possible the requirements analysis will also identify programs required for fault analysis in order to perform necessary maintenance to restore normal system operations.

The most critical step in a software design and development effort is the startup requirements analysis. The misallocation of software requirements occurs when the analysis and definition is split among several functional organizational areas making proper technical coordination nearly impossible. An effective means of avoiding this problem is to assign the total responsibility for determining software requirements to a central system engineering activity.

Concurrently with requirements analysis and definition, a preliminary trial software design for a given system application should be undertaken. This design will yield a functional model for the purpose of requirements validation. Evaluation and trade studies based upon the trial design will result in validating the allocation of both computer and software requirements. Computer resources such as core memory budgets, execution time budgets, and peripheral data storage budgets can be allocated more effectively among software elements by employing a functional design model than without one. These resources allocations must be made to verify that requirements are compatible with operational constraints. Preliminary interface definitions can also be accomplished to ensure compatibility of design requirements among the various computer and software elements.

Computer program requirements are specified in terms of type of program, the instructions to be executed, type of input–output, and frequency of program usage. In addition, parameters may be specified which govern the number, frequency, speed, and priority of instructions to be executed. These later constitute the basis for a *queueing* analysis to determine program control requirements.

When requirements have been compiled, and a functional design model has been constructed, computer processing simulation studies can be run to identify the effect on the processing load of each parameter to be employed. The result is to finalize the requirements allocation and to verify or modify designs of general and/or special processing units sizes and speeds; core memory size and space allocations; input and output speeds; and, peripheral device characteristics.

Computer/Software Design Trade-off Studies

Once the information processing requirements become fairly well identified, along with the initial simulation of preliminary functional designs, a number of trade-off considerations become possible. They are accomplished as a basis for optimizing the overall computer/software design. Listing of trade-off considerations can never be exhaustive, because each design must be derived to uniquely fit a given set of requirements in relation to the intended system operational applications (as reflected in the scenarios). However, some typical trade-off considerations are:

- Automatic *versus* manual control of operations.
- Centralized *versus* distributed control of operations.
- Software *versus* hardware control of equipment operations.
- Analytic *versus* tabular expression of control functions.

- Alphanumeric information display *versus* graphic presentations.
- Information printout *versus* cathode ray tube display.
- Cathode ray tube or electroluminescent display *versus* trans-illuminated displays.
- General purpose *versus* special purpose machine language for programming.
- Complete *versus* selected data processing.
- Dedicated *versus* time-shared processing capability.
- All standard *versus* optimized input and output processes.
- Single processor *versus* parallel redundant processors.
- Rapid access time to mass data storage *versus* slower access at lower cost and increased accuracy.
- Multiplexed *versus* multiple input–ouput channels.
- Permanent *versus* temporary central data storage.

Again, it is emphasized that the list of trade-offs is intended to be illustrative only rather than comprehensive. Specific computer/software design trade-offs to be accomplished will be dependent upon the type, speed, and amount of information to be processed. Modularization of computer and software designs permits great flexibility in tailoring system designs to fit varying information processing needs.

Software Design

Once the computer/software integrated system design has been optimized, detailed design of computer programs can proceed. Computer/software technology has advanced materially not only because of the advent of large-scale-integrated circuitry with resultant increases in speed and accuracy of memory access and information processing, but also because of the modular design of software programs. Modularization of computer programs permits buildup of a library of subroutines possessing a high degree of interface compatibility. Also, because of increased computer capacity, more flexible universal computer languages (e.g., FORTRAN) can be employed rather than being restricted to specific machine languages where improvement in programming efficiency must be gained by increasing logic complexity. Slight gains in programming efficiency of a machine language with a limited number of instruction words involve packing several elements of data into a single word and employing many tricky programming shortcuts. Consequently, such machine language programs are hard to write, check out, modify, and integrate with other interfacing programs.

Use of modular programming employing higher-order languages, taking advantage of flexibility in instruction wording for coding purposes, permits compilation of groups of independent subroutines. A computer program composed of such separate modules of subroutines is easier to write, code. check out, modify, and integrate with other similarly prepared programs than ones that are written in specific machine languages. Without this approach to computer programming design, the increased speed, capacity, and efficiency of modern computer designs would be largely wasted.

In addition to modularity in design of computer programs, there are other important design characteristics from a system engineering viewpoint:

Adaptability to Development Testing. Development test methods should be planned while designing a computer program. To implement this objective, a program should be organized so that test data are easy to obtain. To guide the process, verification test points should be identified in data flowcharts before coding is initiated. Test data can then be related to flowcharts in "desk-checking" a computer program. Output statements causing meaningful data outputs at the identified verification points in a routine or subroutine can be inserted for development testing and later can be removed or made nonoperative. The objective is to try to check out every logic path and arithmetic statement at least once at the lowest subroutine level.

Input–Output Data Isolation. Computer programs should be designed so that control data and test data are easy to introduce and the output data are easy to read and interpret. Discrimination needs to be exercised to select only significant input data and to avoid any permanent recording of voluminous data not of primary interest to the operational user or the test analyst. If the structure of a computer program permits multicase development testing, a method for transmitting data between cases must be provided. When data can be grouped for separate development test cases, a means should be provided for purging data pertinent only to a current test run in order to go on to the next. This prevents irrevocable error in processing one test case precluding the processing of subsequent follow-on cases.

Interface Considerations.   Modular programming is a convenient method for solving many problems in software design and development, but it does not in itself prevent problems when total program level testing is undertaken. Problems can normally occur from two sources:

1. The input–output interfaces or methods of communication between modules, routines, or subroutines were not well defined; and,
2. The modules, routines, or subroutines are not strictly independent, but rather affect each other in complex hard-to-understand ways, particularly in regard to using common computer resources.

Flowcharting

Design of computer programs proceeds by developing flowcharts preparatory to coding. Flowcharts are a means of presenting information and operations in diagrammatic form so that they are easy to visualize and follow. They show the flow of data through an information processing system, the operations to be performed, and the sequence in which they are to be performed. Flowcharts can be misinterpreted because of lack of uniformity in the meaning and use of specific symbols. However, standardized flowchart symbols have generally come into use and when properly utilized should tend to reduce misinterpretations.

A system-level flowchart describes the flow of information through all parts of a system. At the next lower level of detail, a program flowchart describes what takes place in terms of specific inputs, processing operations, decisions, and data outputs. Flowcharting is carried down to subroutines where the programmer can take over from the designer and translate the requirements into a computer language.

The three basic flowchart symbols are for input–output, processing, and decisions tied together by flow direction lines. Figure 8 presents an illustrative program flowchart using these symbols.

The input–output symbol is used to denote any operation of an input–ouput device. Making information available for processing is an input operation. Presenting or recording processed information is an output operation.

The processing symbol is used to represent any kind of processing operation, such as the process of executing a defined operation or group of operations resulting in a change of value, form, or location of information, or in the determination of which alternative flow direction is to be followed.

The decision symbol is used to depict a point in a program which determines a future action, such as a branch to one of two alternate paths. The decision to be made should be clearly stated.

A terminal symbol is used to indicate the point at which the given program originates or terminates.

The flow direction is a basic requirement to represent the direction of processing flow. It is inherent in computer programs that many decision steps are involved. Looping in a program, that is, repeating an instruction sequence, is also a common occurrence. This can lead to complex program flowcharts. It is important, therefore, to draw flowlines clearly and cleanly without crossings.

Descriptive titles contained within symbols should be short but not confusing. For better understanding, the language used in a program flowchart is English, rather than a computer language. Wording is condensed to fit within a symbol without overcrowding. Because of possible ambiguity in meaning, abbreviations are not employed.

The usefulness of flowcharts is enhanced by presenting the flowchart on one-half of a page and the corresponding explana-

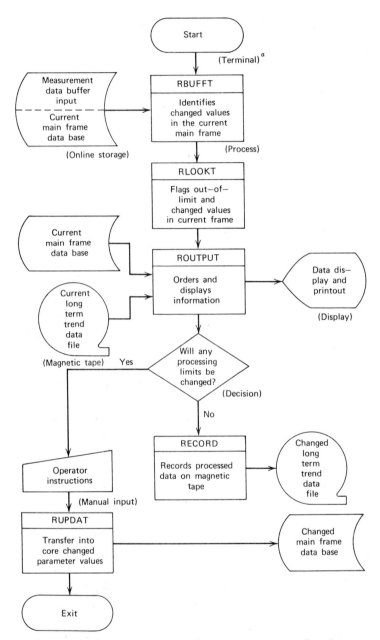

Figure 8. Illustrative annotated computer program flowchart. Note *a:* Symbol meanings are indicated for explanatory purposes. Normally, they would not appear on a working flowchart.

Program name—RPROSTM

Purpose: The measurement data processing program provides the capability for processing equipment operating status data.

Inputs: Measurement data from input buffer.

Outputs: Display of changed measurement values.

          Record of selected values in trend data file.

          Updated data base.

Module name—RBUFFT

Purpose: To empty the measurement data input buffer for data processing and to extract from the current data input only those parameter values which have changed in relation to the current main frame data base.

Inputs: All parameter values from the current main frame data base and from the measurement data input buffer.

          Flags identifying which parameter values are not to be processed.

Outputs: A working table of current values and identifiers of measurement parameters which have changed from those in the main frame data base, excluding those values which have "do-not-process" flags.

Module name—RLOOKT

Purpose: To check the working table of changed parameter values against predetermined numerical measurement limits and to flag those which violate limit checks.

Inputs: Table of changed parameter values from RBUFFT.

          A measurement value processing matrix from the data base.

Outputs: Flagged values which identify parameters violating limit checking or depict status change.

Module name—ROUTPUT

Purpose: To format and route for display all status changes and out-of-limits measurement values.

Inputs: Flagged values from RLOOKT.

          Current main frame data base values for flagged parameters.

          Current trend file data for flagged parameters.

Outputs: Format display messages, item groups and graphs.

Module name—RECORD

Purpose: To make a magnetic copy of processed operating status measurement data.

Inputs: Changed operating status values.

          Equipment operating status and out-of-limits messages.

Outputs: Updated file of selected parameter values on magnetic tape.

Module name—RUPDAT

Purpose: To transfer into the main frame data base the changed parameter values.

Inputs: Working table of changed parameter values.

Outputs: Changed main frame data base.

tory narrative on the other half. This provides a running annotated description of the processing flow as shown in Figure 8.

Flowchart Procedures and Practices

A computer program flowchart is used by the designer to provide
a means of visualizing during the development of a program, the
sequence in which arithmetical and logical operations should
occur, and the relationship of one portion of a program to
another.

In the development stage, a flowchart serves as a means of
experimenting with various approaches for laying out the logic
of the data processing functions. Starting with symbols representing the major functions, the designer first develops the
overall mainline logic by adding blocks to depict input and
output functions, steps for the identification and selection of
stored information, and decision functions.

After the overall mainline logic of the program has been
tentatively established, the primary functions are described in
further detail. The goal is to produce flowcharts which clearly
show all major basic decision points in the program. Such
graphical documentation is used to verify that the program's
procedure satisfies all possible conditions which can arise during
the operations it is designed to control.

Once the procedure is established and appears to be sound,
the detailed program flowcharts define routines or subroutines
and become the basis for coding. The amount of detail will
depend on the purpose of the flowchart. The peculiarities of
machine logic for the computer to be employed may necessitate
changes in program logic. It may, therefore, be necessary to
redraw and reverify a flowchart after coding.

Upon completion of coding, the program documentation is
updated to reflect precisely the finished version to facilitate future
modifications which are bound to become necessary as a result
of testing and the subsequent operational check out of the
procedure. Since it serves as a map of the program sequencing,
the flowchart simplifies the task of modification.

Final documentation of a program should include both the overall mainline logic flowchart and the detailed flowcharts. A mainline logic flowchart promotes understanding of the more detailed flowcharts and also provides an easily understood graphic representation of the sequenced procedures.

Coding from Detailed Flowcharts

Each detailed flowchart of a routine or subroutine and its accompanying annotation is intended to present instructions for coding. The procedure it describes should be self-explanatory in order to be readily understood by any qualified coder. Most procedural descriptions outlive their author's responsibilities for them. Comments describing and summarizing processing requirements, including applicable algorithms, should be placed directly in apposition to the flowchart symbols to which they apply. This information is especially important to those who may inherit the task of modifying the program. Explanations of any peculiar array dimensions, storage limitations, interfaces with other routines, and the like, are also important in this regard.

Coding should be supervised by the programmer who prepared the detailed procedural descriptions. The coding should not be accomplished until after the program and its subroutine requirements have been thoroughly reviewed and verified. The coding and its debugging will end when an error-free program compilation has been made, and the routines cycle successfully.

Software Documentation and Interface Control

As an adjunct to determining requirements and design of software programs, an essential system engineering function is to prepare the computer program development and product specifications, and to exercise software interface control by means of interface control documents.

By the start of a computer program design phase, a complete, well-defined, allocated, and validated set of detailed software requirement specifications should have been prepared. The importance of achieving stability of these requirements before undertaking detailed computer program designs cannot be stressed too often or too strongly.

The first set of specifications to be employed as a basis for defining a system computer program should describe all functional performance requirements, design constraints, and standards to be employed. The role assigned to the computer program in relation to equipment operations should be stressed to delineate the functions it must perform to achieve system output objectives. Specifically, the computer program development specification should provide descriptions of:

1. The overall system operations and design characteristics, its environment, and the role that the specific computer program will be expected to play.
2. The peripheral input–ouput equipment with which the program will be designed to interface.
3. Any existing programs with which the new program will be required to interface.
4. System-level functional flows and function interfaces with associated timing and sequencing requirements.
5. All input data, covering their source, method of insertion, and validity checks. Quantity and timing of the input data and associated limits should be specified. Operator control requirements should also be described.
6. A textual and mathematical description of the processing requirements for each function.
7. All input data, control parameters, and displays to be employed. Method and timing of outputs should be described. Operator output requirements should include type, content, timing, format, and routing of information.
8. Any special data processing requirements, or instructions for special formats to accommodate testing, recording, simulation, necessary procedures, growth requirements, recovery requirements, and special operator support requirements.
9. Test requirements for verifying and certifying program operational validity.

When computer program functional design has been completed and verified as satisfying the performance requirements set forth in the development specification, a product specification is prepared. This specification enables programmers, who are production experts, to take over from the designers. It contains all of the information essential for translating requirements into computer language. The computer program product specification should provide description of:

1. Inputs, outputs, and functions to be performed by each program subroutine and any subroutines common to other programs.
2. Guidelines for mnemonic labeling of all data items.
3. Allocation of memory storage to program subroutines, the executive control routine,* and the data base.
4. Timing and sequencing requirements and equipment constraints used in determining their allocation.
5. Functional flowcharts from the system level to the lowest subroutine covered by the specification.
6. Program interrupts and their effect upon designing the control logic. Each interrupt should be fully described as to source, purpose, type, frequency, and the required response of the executive control.
7. The control logic employed in referencing each routine or subroutine. The details concerning the assignment of priorities and permissible cycle times for each routine or subroutine should be covered. The flowcharts should be annotated to explain the control logic.
8. Any special control features that affect the design of the control logic but which are not a part of the normal operational functions (e.g., loop tests for checking the equipment operating status).

As a computer program design progresses, interfaces become more detailed and the need for interface control becomes critical. Software interface problems are most frequently identified during test and integration. However, it should be planned to solve interface problems long before operational testing by exercising tight control over interfaces as they are defined. To make controls enforceable, it may be necessary to prepare special interface

---

* The *executive control* routine is a routine designed to process and control other routines.

control documents incorporating interface definitions for all designers and programmers to use in conjunction with development and product specifications.

Human Engineering Considerations

In the overall design of a computer/software information processing system, decisions must be made concerning the role of an operator in the system. In general, humans are superior to computers in making decisions in three respects: (a) They are capable of inductive reasoning; (b) They are able to make inferences from one set of conditions to another; and, (c) They are capable of making decisions in situations which they have not previously encountered. In contrast, the logic required for computer operations must be programmed, and, accordingly, must be deductive. As a consequence, computers are superior for performing high-speed information processing and calculations which can be guided by rigidly prescribed logical procedures employing sequentially "branching" decision paths, such as have been described in this chapter.

Assuming that an operator is employed to "close-the-loop" by performing decision-making functions in a computer-driven information processing system, human engineering considerations in the design of the computer programs are very important. Several aspects of software design intimately interface with human capabilities for interpreting displayed computer output information; then reacting to it in a problem-solving context in order to formulate permissible responses; and, finally, composing appropriate input instructions to continue the information processing operation. Assuming further that an operator is properly qualified to understand the information displayed and to determine the correct input instructions in relation to it, proper design of computer programs will facilitate an error-free performance by taking into account that human capabilities impose the kinds of constraints described below:

1. In contrast to computer processing of input information guided by algorithms for selecting alternate paths, human decision making employing standardized procedures is relatively

very slow. Accordingly, information processing for a time-critical high-speed control function, which can be logically programmed, should be automated whenever possible. If, however, decision making cannot be preprogrammed for computer processing, then human decision making is essential, regardless of the time that may be consumed.

2. If reliance must be placed upon human decision making, then computer programs should be devised which will best support the human decision-making role by being designed to present only that information which is relevant to the problem to be solved.

3. Information to be presented for human interpretation should be displayed in as positive and unambiguous manner as possible to facilitate ease of understanding and selection of appropriate responses. To achieve this objective, emphasis should be placed upon designing computer programs to convert output data to simple alphanumeric information displays, preferably in plain English and by easily understood numbers. Information which is different from the last presentation or which indicates an out-of-limits condition should be clearly indicated as such.

4. Human decision-making time increases significantly with increases in the number of alternate choices presented simultaneously. There is a decided advantage, therefore, if the number of alternatives to be presented in a given information output can be kept to a reasonable minimum. This is especially true when the decision-making procedures are not standardized, but are dependent upon the operator's understanding of the principles of operation as a basis for arriving at a course of action.

5. In dealing with information processing requiring a time-critical operator response, it must be realized that as he is required to respond more rapidly ("speed stress"), or at the same rate to a greater number of displayed information "bits" ("load stress"), performance accuracy can be expected to decrease proportionately. Accordingly, when operator decision making is essential the output information stream should be buffered for presentation in an amount and at a rate which is comfortable to the operator. Preferably, he should be able to control the rate of presentation to his own liking.

6. Critical information displays requiring an immediate opera-

tor response should be accompanied by a warning signal, preferably both a flashing visual signal and an auditory signal requiring an operator response to squelch.

Since we are considering the integrated design of a computer/software system in which the operator is an integral functional component, it is important to note in passing that the physical characteristics of displays and controls and their configuration also involve important human engineering considerations. These characteristics interface significantly with software human engineering considerations. For example, it is important that:

1. Displayed information be clearly visible and easily read without inducing confusion as to meaning.
2. Controls for input information should be placed for convenient operation and should be so configured that they can be readily manipulated to facilitate error-free insertion of instructions.
3. The workspace should provide convenient stress-free conditions for task performance.
4. Environmental conditions should be controlled to provide comfortable operator working conditions.

Finally, and not the least important, computer/software system operating instructions and procedures should facilitate the development and maintenance of proficiency in operator task performance by providing adequate instructions on computer operations and the use of the available computer programs for given information processing applications.

Computer Program Testing

Computer program testing from a system engineering viewpoint follows the same basic steps as have been described for hardware items. However, because of the nature of computer programs, there are unique test methods which can be employed to verify their designs.

Although the test methods in some cases modify the coding, they never modify the logic flow through the subroutine being exercised. For this reason, it is valid to use these methods in all development tests except for timing and real-time interfaces. This means that tests designed to verify logic, computations, data handling, and nonreal-time interfaces can be conducted using specially instrumented software. Storage, timing, loading, and real-time interface testing must be performed without the special instrumentation.

Computer program test procedures can be subdivided as follows:

Logic Test Procedures. Logic testing is conducted to ensure that every instruction in the software is executed at least once prior to initial operational application. This category of testing verifies that:

1. Every entry point is exercised.
2. Every exit point is exercised.
3. All error messages are triggered and authenticated.
4. All decision points and branches within a subroutine are executed properly.

Computational Test Procedures. The purpose of computational testing is to verify the accuracy with which software performs operations necessary to achieve quantitative results. Verification of computational adequacy and accuracy is most easily performed at the subroutine level using normal program outputs and supplementary results which summarize intermediate calculations.

Computational testing will provide proof that the subroutine functions properly at its design limits by use of the following kinds of inputs as appropriate:

1. Nominal or expected values.
2. Null data (input data missing).
3. Extreme values (minimum and maximum) including those associated with any options.
4. Out-of-bounds values.

The results obtained from these tests are compared to answers obtained by independent but similar programs, hand calculations using the programmed algorithms, or hand calculations using different but equivalent algorithms.

Data Handling. Although logic and computational testing involve the handling of data, certain additional tests are required to ensure that:

1. Data editing is properly carried out, including limit checking of input data.
2. Input data are obtained from the proper location.
3. Output data are stored in the proper location and format.
4. Data conversions are properly handled.
5. Bad data are properly handled.
6. Data are not lost or destroyed intentionally.
7. Correct input–output devices are properly used.
8. All required data are being handled without reaching saturation and that saturation margins are identified.
9. Only required data are being handled.
10. Data reduction, compression, and/or generation are correctly carried out.

Interface Testing. Interface testing involves demonstrating that groups of subroutines interact properly when executing together. Interface testing will normally involve validating the proper transmission of data between subroutines, modules, and programs. These data flow tests will ideally be a repeat of earlier tests, performed to independently validate the interface coding.

Interface testing should be designed to begin at the subroutine level within a module and eventually include module-to-module, and ultimately program-to-program verification, if more than one program is required within a system. The last tests should be conducted as late as practical since they will require the use of many diverse elements of hardware and software which should be individually checked out. There is a need to conduct some system-level interface tests as early as possible, however, since the cost of correcting interface problems is much lower the earlier they are detected in development testing.

Storage, Timing, and Loading Analyses. Software development tests provide a convenient way to gather statistics concerning the running time required for various subroutines. All subroutines which have specific timing requirements should be clocked for comparison with requirements identified in predesign analyses. Similarly, data can be collected on loading to verify data storage allocations and access speeds.*

Use of Automated Test Methods

Producing error-free software is the final goal of software development. Various automated software test methods can be utilized to assist in design verification. Where in the process these methods can be utilized are illustrated in Figure 9. Briefly, they accomplish:

1. Traceability of requirements to computer program development specifications.
2. Interpretive testing of the correctness of computations, storage references, executive sequences, and other operations.
3. Quantitative measurement of the thoroughness of testing through the successive steps of test planning, test procedures, text execution, and test evaluation.
4. Estimation of program running time and storage requirements for each routine, module, and program.
5. Comparison of program output.
6. Control of test case tapes or card decks.

All development tests are conducted in accordance with a test plan and procedures designed to verify that designs fulfill requirements set forth in computer development specifications. Each test plan should contain a milestone schedule for check-out of the routine or subroutine involved; a specification of the test methods and techniques to be employed; the functional capabilities to be demonstrated; and, a description of each test to be conducted. The use of automated test methods, auxiliary programs, and utility support programs must also be described.

---

*A discussion of computer processing speeds has been deliberately avoided. Suffice it to say that with the advent of LSI and advances in circuit-packaging technology, speeds are measured in terms of millionths of a second (microseconds) and billionths of a second (nanoseconds).

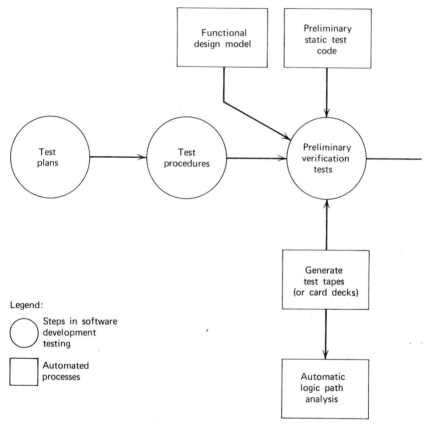

Figure 9. Automated processes that assist in software development testing.

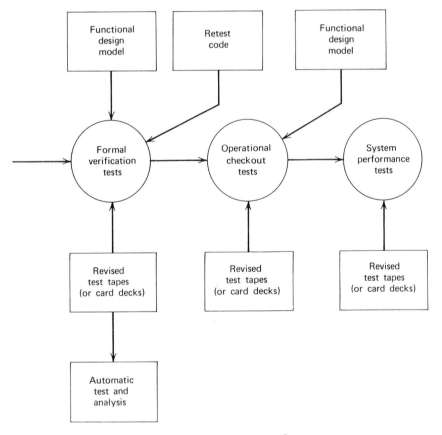

Figure 9. *Continued*

Further, these plans should contain quantitative data indicating the range of values to be supplied on input and, where applicable, the results expected. The plan must relate to the detailed design package upon which the coding is dependent, to ensure that all requirements have been met, and to demonstrate the fact that the actual code within the subroutines or routines is completely exercised. Automated test processes should be able to produce documentation which verifies this completeness.

Unit Test Folders

A unit test folder should be maintained to the lowest level at which the software is to be controlled. Each test folder should contain the following:

1. Detailed design requirements for the subroutines or routines as extracted from the computer program development specification, including applicable design flowcharts.
2. Applicable interface definitions.
3. Development test plans for each routine or subroutine.
4. Source listing and load map for each routine or subroutine.
5. Routine or subroutine flowcharts.
6. Test case results.
7. Check sum for each routine or subroutine used during the execution of the test data.

A cover sheet on the unit test folder should contain the predicted date, actual date, and review date for each milestone in the test plan. These test folders will provide a concise picture of the development process for each routine or subroutine. Of particular importance will be the evidence that each statement of each routine or subroutine has been executed during the course of testing. These folders will constitute a record which can be employed during all follow-on modification programs after initial operational use.

Test Data Catalog

As each test case is constructed, it is given a unique identification number and the test data are stored in a form for computer processing, such as on a magnetic tape. To execute any of the test cases so cataloged, the user simply enters the test number, and a cataloging program searches for it and causes it to be retrieved. Contained within the test case data are comments which document the purpose of the case and the expected results. In many instances, the comments will contain actual answers derived from similar programs used to validate certain options by independent programs.

This test data catalog approach has been proven to be extremely useful whenever retesting is required, such as when adding new capabilities to the program or correcting errors. The user is encouraged to retest since he will have an extensive library of cases from which to choose. The ease with which cases can be added to the library and subsequently executed also promotes the development of new test cases, so the library is constantly being upgraded as the program is modified and improved.

Operational Testing

The primary goal of computer program testing is to demonstrate that the software performs in accordance with specified operational requirements. Consequently, orientation of the operational testing effort is somewhat different from that of development testing in that it is based strictly on requirements documents and has no necessary relation to the software detailed design. As a corollary to this goal, a major objective of operational testing is to uncover those errors not found in the course of development testing. Typically, these will be related to interfaces between hardware and software elements. Problems may also be uncovered when realistic input data are used rather than nominal values. In general, operational testing is the first opportunity to drive the software with a completely external driver. It is normal for such testing to uncover errors not previously found using limited amounts of input data and internal drivers exercising only a limited number of all possible logic paths.

Conclusion

Only the highlights, essential steps in the process, and unique techniques of computer/software system engineering have been covered in this chapter. Detailed design of computers and software is the speciality of computer designers and of software programmers respectively, and is not a responsibility of system engineers. The selection of the material which has been presented would in general answer the question, "As a system engineer, what does one need to know about computer/software design and development to deal with system engineering problems involving the application of automated information processing technologies?"

8

System Engineering Management— Dealing with Intangibles

Obtaining a Proper Balance in Management Control

How best to manage a system engineering effort has puzzled many well informed system project managers. It seems as though every possible attack upon a problem has been tried—and found wanting in some important respects. From a customer's standpoint, approaches to system engineering management have ranged all the way from granting complete permissiveness to a system developer by saying, "Here is the problem, you give us the solution," to attempting to levy rigorously prescribed procedures upon a contractor which demand that, "You carry out the instructions we give you, and we will inspect you at frequent intervals to ensure that you are carrying out our detailed procedures." As is usually the case in trying to attain some management objective, neither extreme is satisfactory, either from a buyer's or a seller's standpoint.

Creating and producing a performance- and cost-effective system must be a mutual undertaking. To use a well known cliché, a system must marry a customer's desires with a producer's technical ability to satisfy them. Overcontrol on the part of the customer will be costly and will not increase the producer's available inventive or creative talent. Undercontrol is also costly, because it is most likely to result in an unsatisfactory initial product which will require an inordinate number of design modifications to fulfill a customer's expectations. Consequently, as in all situations which involve two or more human beings attempting to reach an agreement upon complex issues, there must be a true dialogue between user and producer.

The ground rules for achieving mutually understood dialogue are not easy to state. They are even more difficult to implement once they are agreed upon, because effective system engineering management involves understanding and acting upon intangibles. The difference between a "good" and a "bad" system is not readily perceptible. As we have previously discussed, the evaluation is always made in relation to values which are by their very nature subjective. There are, however, some criteria, principles, and practices that can be described which, when properly applied by knowledgeable individuals, can lead to successful system engineering management. Before attempting to describe them, it is necessary to reemphasize that a general understanding and feeling for what constitutes a "system" is of paramount importance and above all else such an understanding is essential before one should even attempt to undertake to "manage" a system design and development effort. Without a system perspective, project management just naturally is preoccupied with hardware end item design and development at the expense of system performance goals.

Some Important Definitions

If one is to achieve intelligent management of anything as intangible as system engineering, it is important to have a clear understanding of all aspects of the process of conceiving and creating a system design. There has been much confusion about certain terms and, consequently, much overlapping in their usage. It is necessary, therefore, before proceeding further, to define some key terms.

System Analysis. This term is most meaningfully employed to describe the class of studies which are concerned with determining whether all facets of a complex system can or will contribute to the achievement of its intended operational purpose with desired economy of design and support. The purpose of system analysis is to ensure that the "big picture" is clearly understood, that a clear-cut relationship is established between the need and the design and that achievement is related to purpose. An essential feature of system analysis is the accomplishment of trade-off studies to establish competitiveness in selection of system elements for alternative system design approaches, and to point up possible weaknesses in the design approaches under consideration. Also, in the process, it is important to establish the sensitivity of each system parameter, both internally in terms of the interplay of system elements, and externally as they relate to the intended use environment.

System Engineering. This term refers to the process of translating operational requirements into engineering functional requirements and subsequently expanding these functional requirements into detailed equipment and service end item design requirements. This process involves analyzing system performance requirements, performing system-level trade-off studies, synthesizing alternative system design solutions by employing various combinations of equipment and service end items, and, finally, selecting the preferred candidate configuration which best meets system performance- and cost-effectiveness criteria.

System Engineering Management. This activity is concerned with monitoring and controlling the process of deriving and producing a coherent system design to achieve stated operational requirements. It involves exercising an overview of the engineering design and development process to ensure that the interrelated roles of all necessary design disciplines and engineering functional areas are effectively utilized for satisfying total design requirements. Again, total system design requirements means the proper mix of hardware, computer programs, facilities, personnel training, procedures, and logistics support as required to achieve an integrated system performance- and cost-effective system design. Further, the required system and end item design and development effort must be interrelated with the other system life-cycle requirements for fabrication, installation and check-out, test and evaluation, deployment, production, modifications, maintenance, logistics support, and phaseout (planned obsolescence). System engineering management is a function which must be exercised throughout all phases of a system life cycle if system design integrity is to be ensured.

The Meaning of Management

In view of the widespread use and the variety of meanings attributed to the word *management* (as to the word *system* which we have already defined), it is important to clarify its meaning. Management concerns people. It basically consists of planning, organizing, selecting, directing, motivating, controlling, supporting, and evaluating the performance of people acting both as individuals and in concert as a team. Consequently, the only basis of effective management is *communication*. By management communication we mean a *dialogue* between a higher level of decision-making responsibility and a lower one. For system engineering management, without thorough training and understanding of the specialized language (data bits) used for transmitting technical messages between various levels of system management there can be no dialogue, but only parallel monologues. The need for effective system engineering communication and understanding is between: (a) system analysis and system engineering; (b) system management and system engineering; and, (c) system engineering and hardware engineering, including the associated support functions.

There can be no management control between these respective levels of discourse until there is a mutual understanding and meaningful exchange of viewpoints between responsible *individuals* at each level. No amount of compilation of information on arbitrarily prescribed forms and subsequent submission of data reports as a "management control" device can be effective without this person-to-person contact motivated by a mutual attitude of understanding and desire to solve shared problems.

System Management Considerations

The major parameters of system management concern project technical performance, cost, and schedule planning and control. There invariably is a "cat and mouse" game between a buyer and producer concerning the extent and the level of detail to which a formal data reporting system will be applied. Most customers tend to like as much continuous visibility and traceability as they can achieve in accounting for how their money is being spent in relation to what is being achieved. Most system and equipment contractors prefer to have a minimum of formal control over how they choose to commit available funds to achieve specified performance goals.

All planned management controls inherently have a built-in threat of disciplinary action. It is human nature to try to hide one's real or imagined faults to escape the possibility of punishment. Technically trained people are no exception to this generalization. Theoretically, effective system engineering management is supposed to provide identification and visibility of "high risk" areas in undertaking system design and development projects. The fact that a project schedule may be established and cost controls may be placed into effect which are unrealistic in the face of the identification of certain "high risk" areas only focuses attention upon any failure to attain performance goals, even though their occurrence was predicted with a high degree of certainty. The resulting emphasis psychologically is more likely to be on the "failure" rather than on the fact that attainment of a desired technical achievement by a given date had a high degree of improbability on the basis of technological considerations. Such considerations are indeed intangible and hard to depict convincingly within the usual management framework. A "failure" to meet a specific performance goal on a given scheduled date is readily demonstrable and, politically, can be made to appear dramatic for attention-getting purposes. The responsible technical managers are, therefore, exposed to scrutiny by their project management overlords and thus psychologically react as though they will be culpable and potentially subject to punishment. It is entirely natural, therefore, to expect scientific and engineering personnel who are to be responsible for undertaking work in a "high risk" area to try to avoid tight cost and schedule controls and to fight for as much time and budget as they can obtain, prior to initiating any effort.

Although system engineering can function effectively only in a well-structured hierarchy of authority, it is always possible that the overall system project management control of technical performance, cost, and schedule will rest in the hands of someone who acts in a most uninformed and arbitrary manner. Sound system management principles and practices can be effectively applied to system design and development efforts only to the extent that both parties—management and technical personnel, or customer and producer—realistically understand any risks involved and are objective in their evaluation of potential success or failure of technical performance. We will have more to say about risk assessment and technical performance measurement in Chapter 9, but first there are some overall system management techniques to be discussed which are important because of their direct impact upon the effectiveness of the system engineering management process.

Use of Work Breakdown Structure

The first step in exercising system management from a cost and scheduled planning standpoint and as a basis for structuring technical performance criteria is to define the project work packages. A *work package* is usually defined as the delineation of work required to complete a specific job with objective indicators which define the start and completion of the task to be accomplished. The work breakdown structure is developed to the lowest level work package subdivision which represents an identifiable work effort in relation to a given hardware or software product, such as designing a component; producing a set of data; preparing operating and maintenance procedures; or, rendering a service such as training or conducting systems test and evaluation. Further, to be effectively controlled, the overall responsibility for the actual performance of the tasks comprising a work package must be limited to a single operating level organization. The work package is the element to which technical performance, cost, and schedule controls are applied. In the context of planning an integrated system equipment and service end item design and development effort, it attempts to create something tangible out of the intangibles.

Definition of work packages for a new system design and development effort is more tenuous than for a well-defined hardware production effort. In the former case the effort is concerned with attaining an *objective* rather than producing a physical object as in the latter case. Furthermore, in pursuing a research and development objective, there is seldom any comparable historical experience which can be used as a guide in estimating how long a given work package task will take. It is necessary to define the work to be performed in terms of a "level of effort" to be expended until such time as the objectives of the work package have been attained. It is, therefore, necessary to schedule periodic reviews of the work in progress and, as a consequence, to reestimate the amount of manpower to be applied to the task in the ensuing time periods and to adjust project schedules and budgets accordingly.

Objectives of Technical Performance Measurement

A technological breakthrough in a system design and development effort can expedite the attainment of performance goals, and may even result in a cost saving. To encourage a contractor to seek such breakthroughs, contracting agencies have attempted to employ various "incentive fee" arrangements to reward delivery of high-performance products ahead of planned schedules and to penalize late deliveries, increased costs, and failure to meet contractual performance goals. Realistically, technical managers will try to obtain as favorable a position as possible in regard to delivery schedules and cost estimates at the time his contract is negotiated, especially if there is an incentive fee provision tied to them. To avoid penalties, technical performance measurement techniques must be sensitive enough to predict potential failure in relation to achieving a contractual performance requirement in sufficient time to seek customer agreement to a schedule slippage and additional funding, or customer acceptance of a modified performance goal because of clearly recognized technical difficulties not attributable to purely management deficiencies. Without such a procedure, any deficiencies may be revealed only at the scheduled time for product "buy off" by the customer, and then it is too late to expect him to be lenient in extending schedules or agreeing to increased costs.

Considering the increasing costs due to the sophisticated technology employed in creating complex systems coupled with the annual monetary inflation rate, it seems reasonable to assume that a customer, especially a government procurement agency, has a responsibility and a contractor has an obligation to account for, report, and be evaluated on how wisely funds are expended in a system design and development project. In addition to the rather obvious and direct impact of such project management cost and schedule control on the system engineering effort, the related technical performance measurement is generally considered to be a direct responsibility of system engineering management. It constitutes, therefore, a major interface between system engineering management and project management.

The "name of the game" for system project management is to exercise cost and schedule control in relation to attaining technical performance objectives. Interrelating these parameters from a system management standpoint is not easy to do, let alone controlling their interactions. In Chapter 9 we will explore further the role of technical performance measurement in relation to cost and schedule control, and the responsibility of system engineering management for monitoring the progressive achievement of defined technical objectives within project cost and schedule constraints.

Role of System Engineering Management

System engineering management properly bridges the gap between system project management and the detailed end item engineering and production effort. System project management must be primarily concerned with attaining performance goals within any established cost and schedule constraints. Since, by definition, we are concerned with complex man–machine systems, this in itself is a major effort. On the other hand, as we have been emphasizing, a system will not just naturally evolve all by itself by assigning discrete equipment items to be designed simultaneously by hardware-oriented engineers. Someone must make a deliberate and conscious effort to initially synthesize an integrated system design from which the requirements of the various subordinate system elements, subsystems, or component end items can then be derived.

Attempts to devise formalized "system engineering management procedures" have only resulted in the development of a folklore that system design and development can be regulated so that it will proceed in an orderly logic-tight manner by progressively: (1) defining system requirements; (2) performing trade-off studies on alternative design solutions; (3) selecting a preferred solution; and, (4) deriving and testing end item equipment designs which fulfill the original system performance requirements. It is hoped that the fallacy of this folklore concept has been thoroughly exposed by the time the reader has reached this point in our discussion. The most practical and apparently feasible device for a system-oriented project manager to employ for communicating effectively with his project staff is a series of well-chosen "checklist" questions. System engineering requirements differ for the various steps in the life cycle of system design and development. Accordingly, the presentation of appropriate system engineering questions will be grouped in relation to the major steps of system concept formulation, system design and definition and engineering design and development.

Concept Formulation

Any system project must start with the recognition and definition of a need. Such a need may arise in a number of ways. A customer may feel he lacks some capability he desires to have. He may have experienced some performance deficiencies in trying to use existing systems. Or the availability of some new technology may induce a desire on his part to be able to use it. In any event, the system engineer's job is not necessarily to question the need, but primarily to examine how best to fulfill it through a system design and development program. As a first step, he should thoroughly acquaint himself with the objectives of the project by seeking answers to such questions as:

· How does the customer describe the basic need for system?
· How did he establish this need?
· What are his financial resources for supporting the design and development effort?
· What is his established need date for operational deployment of the system? How urgent is this date?
· How strongly does he feel he must have the desired system capability?
· How firmly established are the technological capabilities upon which attainment of system performance capabilities are dependent?
· Has the customer developed a scenario that sets forth conceptually how he expects to employ and support the proposed system in the context of its intended use environment?

As system analysis proceeds, and a first iteration of the system performance requirements is made in relation to mission objectives, it will become possible to consider and to begin furnishing answers to such questions as:

· What are the basic objectives of the proposed system design and development effort? Has a clear description been developed that sets forth the intended application of the proposed system design and development effort?
· What are the various operating modes to be encountered and what is the rank order and relative importance of the mission objectives in terms of the desirability of attaining the required design capability to meet the performance requirements for each mode?
· What are the installation and deployment constraints for each mode of operation with regard to such factors as space, size, weight, power, hardness, operability, and so on?
· Do the cost and schedule constraints appear reasonable in rela-

tion to the research and development requirements for solving the known technological uncertainties? If not, what adjustments should be made?

- What trade-off studies have been made to establish the sensitivity of each system performance parameter? In relation to each other? In relation to intended use environment? In relation to cost and schedule constraints?
- Have the "elements" of the system been appropriately identified and adequately defined, their interfaces described? Which of the elements require specific design and development effort? What is the estimated technical risk involved in the development?
- Is the proposed work breakdown structure realistic in terms of the required design and development effort for each of the system elements?

As successive analysis iterations are accomplished, it should be possible to answer such questions as:

- Have the necessary technologies and design disciplines required to design and develop the system been identified? Has the specific role that each is to play been described?
- Are the technological capabilities fully available as a basis for proceeding with immediate system design and development? If not, what research and development is required to provide a given technological capability prior to proceeding with full-scale system design and development?
- Have the operations, maintenance, and support functional requirements been adequately described as a basis for specifying system and system end item capability, availability, reliability, maintainability, transportability, supportability, and other appropriate design characteristics?
- Have the system performance-effectiveness model criteria been developed as a basis for planning a system design analysis and test program for progressive evaluation of system design and development?
- What is the estimated cost effectiveness of the proposed system design approach in terms of its projected operational deployment and support over the specified period of useful life?
- Has a system specification been written setting forth complete

qualitative and quantitative performance and design requirements as a basis for proceeding with the definition of performance requirements for the various prime mission and direct
support equipments and service end items?
- Have the technical program plans been fully developed to
identify and define the requirements for *system* design, development, fabrication, installation and check-out, test and evaluation, deployment, production, maintenance, logistics support,
and eventual phase out?
- Do the cost and schedule estimates appear to be reasonable?
- Have the work statements for the definition and/or design
and development phases been prepared in a manner which
properly directs an integrated system engineering and design
effort? Without redundant parallel application of several design
disciplines as an ancillary effort not in the mainstream of the
engineering design and development process?

System Design and Definition

From concept formulation, a system project may move into a
formal design and definition phase, or such a step may be co-
alesced with the engineering design and development effort. In
any event, there are appropriate system engineering management
actions to be taken. Normally, the system engineering management effort now becomes more regulatory and, generally speaking, is placed in a position of directly monitoring and controlling the total engineering and design effort from the standpoint of ensuring the realization of an integrated system in terms
of fulfilling the system performance specification requirements.
For convenience, the checklist questions have been divided into
those pertaining to: *Basic Planning and Control Requirements;
The Analysis and Study Process;* and, *Documentation.*

Basic Planning and Control Requirements.
- Has the documentation resulting from concept formulation
studies been reviewed and analyzed for completeness and
coherency? Has the information been obtained or developed
for any omissions in the concept formulation documentation?

- Has the project work breakdown structure as furnished for the system design and definition been further detailed to identify work package tasks to be completed for analyses and studies as required to derive a *total* system design and technical development program package?
- Has a system engineering management plan for the integration of system design and definition studies been prepared and promulgated for the guidance and necessary compliance of all who are to participate in the effort?
- Have initial end item equipment specifications and technical program plans for the follow-on engineering design and development been prepared to incorporate the baseline *project requirements* as derived from the concept formulation documentation? Have these materials been made available to all personnel who are to participate in the system design and definition studies so that their efforts can be properly channeled into producing necessary and essential information and to reduce the amount of wasted effort?

The Analysis and Study Process.
- Have the operations, maintenance and support plans for the prime mission equipment been developed in sufficient detail so that they constitute adequate design criteria for use in deriving the performance/design requirements for the various prime mission and direct support equipments and service end items?
- Have the system performance-effectiveness criteria or models been sufficiently developed so that a proper basis is furnished for evaluating candidate end item designs? Are the system sensitive parameters properly identified so that alternative design approaches of interest can be assessed in terms of estimated mission performance capability and availability?
- Have the steps in the prime mission equipment operational employment sequence been defined to the level of understanding necessary to define the performance/design requirements of its component equipment and service end items, including any directly associated operational support equipment, and operator personnel requirements?
- What are the alternative end item design approaches which will satisfy the overall system performance and design require-

ments? Have the alternative design approaches been thoroughly evaluated in terms of their relative system performance- and cost-effectiveness?

- For the favored candidate alternative prime mission equipment performance/design approaches, what will each require in the way of support equipment, facilities, computer programs, technical manuals, personnel training, and logistics support in order to achieve the required level of availability from the maintainability standpoint? What will be their relative cost effectiveness for the expected useful life of the system?

- Has a system design requirements review been programmed at which the prime mission equipment design approach will be selected in relation to best satisfying the stated user requirements? And the system performance- and cost-effectiveness criteria or models?

- Following the system design requirements review for selecting the prime mission equipment design approach, have responsibilities been assigned for preparing all inputs as will be required to initiate detailed prime mission equipment design and procurement activities in the follow-on engineering design and development? Do these include the preparation of detailed design criteria covering capability, availability, reliability, maintainability, safety, survivability, vulnerability, supportability, and other appropriate design disciplines within the framework of the established system performance-effectiveness criteria or models? Do they cover design studies for deriving performance/design requirements for equipment components? Schematics and other interface definitions for the functional interrelationships of system segments or equipment subsystems (or component equipments)? And, with the personnel who will operate and control them?

- Following the selection of a design approach for the prime mission equipment, has a functional requirements analysis been accomplished as a basis for determining the direct operational support equipment requirements? The maintenance equipment requirements? The computer program and technical manual requirements? The spares and repair parts provisioning requirements?

- Has a system design requirements review been programmed at which the integrated direct operational support system design approach will be selected?
- Following the system design requirements review for selecting the integrated direct operational support system design approach, have responsibilities been assigned for preparing all inputs for performance/design requirements specifications as may be required to initiate detailed support equipment, facilities, and service end item design and procurement activities in the follow-on engineering design and development? Do these include the preparation of detailed design criteria covering capability, availability, reliability, maintainability, and other appropriate design disciplines within the framework of the established system performance-effectiveness criteria or models? Do they include design studies for deriving performance/design requirements for equipment, facilities, personnel training, computer programs, technical manuals, and logistics support components? Schematics and other interface definitions for the functional interrelationships of direct operational support system components with the prime mission equipment and among the components of the support system and associated facilities? And, with the personnel who will employ them?
- Following completion of the definition of the prime mission and the direct operational support equipments and the respective end item performance/design requirements for each, has the "make-or-buy" decision been made?
- Has the project work breakdown structure been extended to cover the work package requirements for preliminary and detailed design and related technical program activities for the follow-on engineering design and development?
- Have responsibilities been assigned for preparing inputs for the technical program plans for engineering design and development covering all items in the project work breakdown structure in regard to achieving design, development, fabrication, assembly, integration, test and evaluation, installation and check-out, and development of the system?
- Has a detailed analysis been accomplished to define the system

engineering tasks for engineering design and development to include: A description of studies and analyses to be performed in conjunction with deriving detailed designs; methods of control; distribution and use of system engineering documentation; the means of integrating the efforts of all applicable design disciplines such as reliability, maintainability, safety, human engineering, survivability/vulnerability, supportability, and any others; the support of test programs; the procedures for conduct of requirements and design reviews; and, the accomplishment of scheduled technical performance measurements and risk analyses to assess the design and development progress and associated costs in relation to planned project technical performance, cost, and schedule goals?

Documentation.

- Has a project work breakdown structure been prepared as a firm basis for design and procurement activities to be undertaken in engineering design and development?
- Have performance and design requirements specifications been prepared for all prime mission and direct operational support equipments, computer programs, and facilities?
- Have all necessary technical program plans, including a system engineering management plan, been prepared to describe how all equipment and service end items are to be designed, developed, fabricated, assembled, integrated, tested and evaluated in relation to meeting end item and system performance/ design requirement specifications?
- Has a work statement for engineering design and development been prepared which provides that the integrated system design and engineering effort will be continued in the in-line mainstream engineering role of specifying, monitoring, and evaluating the detailed design and development activities to ensure the production and test of a complete system in relation to satisfying the stated user requirements?
- Has a summary report been prepared setting forth the findings of the system design and engineering effort to show that the selected system and system end item design approaches have been optimized both in terms of estimated overall system performance- and cost-effectiveness?

Engineering Design and Development

The system engineering management effort in relation to engineering design and development is concerned with monitoring and controlling detailed equipment and service end item design and development efforts to ensure that their resultant integrated performance and design characteristics will fulfill specified system performance requirements. Checklist questions have been grouped under the following headings: *Basic Planning and Control Requirements; Monitoring and Controlling Detailed Design and Development Activities;* and, *Documentation.*

Basic Planning and Control Requirements.

- Has the documentation resulting from system design and definition studies been reviewed and analyzed for completeness and coherence? Has the information been obtained or developed for any omissions in the documentation?
- Has the project work breakdown structure been prepared to describe work package tasks for all equipment and service end item design, development, production, and test activities?
- Has a system engineering management plan for the technical monitoring, control, and performance measurement of the total engineering effort been prepared and promulgated for the guidance and compliance of all who are to participate in the design and development effort?
- Has the interaction of the system engineering effort with the system management technical performance, cost, and schedule control procedures been clearly identified and an integrated working relationship established?
- Has the interaction of the system engineering effort with the configuration management control activities been clearly delineated?
- Have the work statements for the various prime mission and direct operational support equipment, computer programs, and facilities design and development activities been prepared and reviewed to ensure that the detailed design and development effort will be performed strictly in response to applicable performance/design requirements specifications? And to ensure that any desired deviations or changes will be justified and

processed through the configuration management control activity for system engineering evaluation and approval and preparation of specification changes prior to their implementation as a design requirements change?

- Has the system engineering activity been organized and staffed by properly qualified personnel so that it is capable of giving technical direction to the various activities responsible for the detailed design and development of prime mission and direct operational support equipment, facilities, computer programs, technical manuals, personnel training, and logistics support? Does it have properly qualified representatives of all appropriate technologies, design disciplines and management sciences who have been indoctrinated in the objectives and application of system design and engineering methods and techniques for producing an integrated system performance as an end product of engineering design and development? In the organizational structure, is system engineering management a central, in-line mainstream function with direct surveillance of the detailed design and engineering effort, interfacing with the system management superstructure for project performance, cost, and schedule control?

Monitoring and Controlling Detailed Design and Development Activities.

- Have performance/design requirements specifications and/or work statements been prepared for each equipment or service end item work package identified on the project work breakdown structure? Have performance/design requirements briefings or reviews been held with each activity charged with implementing (or proposing the implementation of) the specifications and/or work statements, so that a clear mutual understanding has been reached?
- Have appropriate preliminary and critical design reviews been programmed at which proposed equipment and service end item preliminary and detail designs will be reviewed and evaluated in relation to compliance with its governing performance/design requirements specification and/or work state-

ment? And in relation to their integration into the total system design in compliance with the requirements of the system performance/design requirements specification?

- Are all proposed equipment and service end item design changes requiring a change to a performance/design requirement specification and/or work statement being processed through system engineering for evaluation of their impact upon the performance/design requirements for other elements of the system?
- Has an integrated test and evaluation program been devised to progressively qualify and verify detailed equipment and service end item performance/design specification and/or work statement compliance and, when integrated, to demonstrate attainment of system performance/design requirements? Does this program properly integrate all phases of testing such as simulation, breadboard/brassboard, engineering model, qualification model, and operational system demonstration tests? Has a central point of control for data processing and evaluation activity been established to assess whether analytical data and test results verify attainment of system performance- and cost-effectiveness criteria and standards?

Documentation.

- Have detailed design (product) specifications been prepared in preliminary form for use in critical design reviews? In final form for use in first article configuration reviews?
- Have a system test plan and procedures been prepared to describe how the verification of all of the performance/design requirements contained in the end item and the system performance/design requirements specifications will be accomplished?
- Has a data collection, reduction, and analysis plan been prepared to show how the achievement of system performance- and cost-effectiveness standards will be demonstrated? Based on system test data, has an equipment modification plan been prepared as a basis for introducing design changes and accomplishing equipment retrofits?

Operational Deployment

There is a need for system engineering management as long as there are requirements for modifying a system. It is not unusual—in fact it is generally normal—that as a consequence of operational use, deficiencies in a system design or operational procedures will be revealed. Introducing a change and retrofitting all systems which have been deployed is an intricate process and requires careful planning and coordination to accomplish. Since each modification project will be unique, suggesting a standard set of checklist items is not feasible. Some of the questions listed for engineering design and development will be appropriate in relation to a design modification project. However, they need to be selected and appropriately modified to meet the requirements for a given modification project.

Some Pitfalls in System Engineering Management

The kinds of things which are done in the name of system engineering management but which have been proven to have little impact in shaping a system design are too numerous to try to enumerate and describe. Generally, in relation to complex man–machine systems, any attempt to devise a formidable data compilation and management procedure is doomed to failure. Any objective study of the real decision-making process for a given system design and development project will reveal that decisions are made in a direct fashion, and in nearly all cases will involve an informal face-to-face contact by responsible project management personnel communicating freely on a person-to-person basis. The transfer of technical information is influenced by human factor considerations which are alien to formal proceduralized approaches devised to record and transmit massive amounts of detailed technical information, often by computerized means, down to the finest degree of minutiae. Only a very tiny fraction of such data is ever employed for technical decision making. Prior experience is the main determinant of technical choice of a given system design approach. From the decision-maker's viewpoint, technical solutions are either initially obvious from the inherent nature of the problem, or they are obvious to him once they are conceived in the course of his problem-solving behavior. No amount of analyses or trade-off studies recorded as "system engineering data" is going to automatically influence or change his decision which is dependent upon his personal understanding and analysis of the problem. If he chooses to employ some of the techniques we have been discussing, then they become effective because they become a part of his analytical capability. Failure to understand this basic human factor consideration has led to a whole host of formalized schemes to gather and record "pertinent facts," to conduct "formal analyses," and to condense and present the results as "system requirements" to presumably perceptive technical managers who, as a consequence, will be powerfully persuaded to alter their decisions on the basis of the results of "systematic investigation." By making a major fetish of formalized system engineering management procedures over the past few years, their instigators have only succeeded in establishing a body of folklore about their supposed utility which places them in the realm of occultism. As such, system project management personnel who are responsible for the real technical decision making have been unalterably alienated from ever being associated with or a party to the elaborate cumbersome schemes for formalizing, and even computerizing, the decision-making process.

The only truly effective system engineering management has to be exercised by knowledgeable, perceptive, system-oriented technical managers working directly within the in-line mainstream design and development effort, answering decision-demanding questions on a day-by-day, person-to-person basis. We cannot possibly overemphasize this fact. An effective system engineering manager must be on hand when a problem happens and he must deal with it immediately if he desires the opportunity to contribute a system-oriented solution. If he waits for some system engineering data scheme with its natural built-in time delay to present the problem and a recommended solution, it will simply be a case where all he can do is to reflect that, "I wish I had been consulted when that problem occurred."

9

Exercising System Engineering Control— The Role of Risk Assessment and Technical Performance Measurement

Management Schemes That Don't Work—A Recapitulation

In seeking to identify and describe the natural approach to obtaining performance- and cost-effective system designs, we have reviewed and discussed a number of techniques which, under certain conditions, can serve a useful purpose. In doing so, we have been attempting to describe and explain the *total environment* for effective system engineering. The need for "visibility" and "accountability" for decision making in the complex organizational structure required to create the type of large-scale complex combinations of hardware, facilities, software, procedures, personnel training, and logistics support which makes system engineering necessary has led, as we have discussed previously, to the concoction of formal formidable management control procedures. As an overall generalization, it must be concluded that they don't work. The basic reason is that human beings adapt to externally imposed rules of behavior by learning how to circumvent them— as long as they can avoid punishment in doing so. Also, for scientific discovery and technological invention, permissiveness is a necessary social condition. Unfortunately, the environment for system design and development efforts by its very nature generally involves pressure cf a tight time schedule, limitations in economic resources, and a sense of urgency for obtaining a usable product, usually requiring the attainment of significant technological advances. Response to these pressures has resulted in a style of system management which has bought large quantities of paper studies prior to committing funds to buy hardware. One study of this problem has compiled the following comparative costs for military aircraft procurement:

Time period	Design study costs	Typical aircraft development prototype costs
1903		$1K (Wright Bros.)
1920's		$10K
1930's		$600K
1940's		$1M–$5M
1950's	$100K–$1M	$10M–$20M
1960's	$10M–$20M	$200M[a] (B-70)

[a] Cost of one of two development models.

With ten to twenty million dollars, as these figures show, being spent in the 1960–1970 decade on design studies for military aircraft systems, it is only natural to expect that a considerable portion of the cost of system acquisition will be expended on management and engineering control systems *per se.* As a consequence of the style which has evolved in government contracting for large-scale defense systems since about 1950, the requirements for massive management and engineering data outputs and documentation prior to and concurrent with system hardware design and development have involved a process of organic growth in both customer and producer system project management organizations. Unfortunately, from the standpoint of furthering an understanding and utilization of the real-life dynamics of system design and development, this growth has resulted in an increase of organizational units in both customer and producer system project management offices dedicated to developing and refining specialized management and engineering data generation and reporting techniques. This growth has also resulted in:

1. An increase in the number of project management communication channels with a corresponding increase in manpower concentrated in system project management organizations.
2. Institutionalization of the project management elements devoted to carrying out the specialized communications so that their roles have become ends in themselves and, therefore, motivated to ensure self-perpetuation.
3. Compartmentalization of management efforts separated from the mainstream of the hardware engineering design and development effort so that, in reality, management techniques generally have little to do with, and, at best, are a minor influence upon, the design and development of system end item equipment and supporting services.

The attempts to develop and promulgate formal system engineering management procedures, when viewed in proper perspective, must be categorized as one of the specialized system management "cults." As was pointed out in the last chapter, no matter how embellished system management becomes with paper analyses, the real decision making which shapes the character of system and end item designs is achieved via person-to-person contacts. A small, close-knit system design and development team with appropriate talents and exercising respected leadership over specialized design groups is still the only really effective way for producing an end product with desired performance capabilities. Does this mean that some means of formal management system, including system engineering management, is unnecessary? The answer is "No." However, the identification of an effective way to achieve such control should not be sought in the direction of formal, formidable, massive documentation. It does, however, reside in the direction of creating a *total environment* which is conducive to the emergence and effective utilization of creative and inventive talents oriented toward achieving a *system approach* with a minimum of management encumbrances. Such an environment would keep the work of the various "cults" concerned with specialized management techniques and also that of the various design discipline "ilities" to a minimum essential for ensuring the appropriate selection and application of basic technologies and design disciplines required to produce a desired product performance capability. Elaborate formalized schemes for management and engineering control of the ongoing design and development process are unnecessary in such an environment.

Contrary to the system engineering management philosophy expressed above, beginning with about 1960, there have been several attempts to establish prescribed procedures for conducting system engineering for government contracts. The result has been an admixture of basic engineering and technical management procedures existing in juxtaposition, but which have never achieved amalgamation. The primary reason why they haven't been effectively married is that the so-called system engineering management procedures have generally been superimposed upon already established engineering management practices. The fact that the attempts to introduce system engineering procedures have been made by promulgating them as a "forcing function" upon the design and development process has had the effect of placing them in the role of a management control system. It is human nature to accept control systems with suspicion and an attitude of unwilling compliance. It becomes a "game" to see how far one can go in circumventing management control. Consequently, control procedures are most apt to be viewed as just an organizational constraint which, as past experience has more often than not shown, will result in the "accusing finger" being pointed at some middle or lower level manager in case a deficiency develops which attracts the attention of higher management.

The accepted general principle of management control is to pass directions "down through channels" and to devise the work breakdown structure so that the source of problems, errors, or defects in carrying out directives can be quickly pinpointed. This means that in the event that trouble develops, an effective management control system will permit a quick descent by higher management to the working level. This "management by exception" technique generally means that the working level supervision never hears from higher management as long as everything is going well. Consequently, management control schemes designed to provide "visibility" and "traceability" may be viewed only as offering potential pitfalls to catch the unwary and thus to be subverted if at all possible. Besides, the working level people see only a whole group of staff people who do not have to carry out the line responsibility for making design and development decision being in a position to say how a particular management effort is to be carried out and to police the quality of the data which come out of it. Seeing the buildup of staff manpower devoted to "paper work" functions usually has a negative effect upon the line engineering organization. The reaction generally is to take a do-nothing attitude such as, "Let the project staff handle the system engineering paper work. Let them find the design problems and flaws if they can. We are not going to publicize our weaknesses just for their benefit." As a consequence, functional design groups in an engineering organization generally adopt a very defensive parochial viewpoint. They become concerned only with accomplishing some relatively small piece of a system and with successfully passing the prescribed quality control inspection in order to complete a given work assignment. Fitting all the separate pieces together and making them work as a system become someone else's problem. After buy-off of their particular work package unit, they will be willing to fix any performance deficiencies which may show up subsequently, provided there is additional money to cover the work involved.

In the face of such real-life circumstances as we have been describing, how is system management control ever achieved? Obviously, it must be done by identifying and capitalizing upon what we choose to refer to as following a "natural approach." It cannot be accomplished by inventing and attempting to apply management control "gimmicks." To be characterized as a natural approach, a system management control procedure must be one that the separate design groups can readily accept and implement in their own interest as well as that of the higher-order "system approach." There are two interrelated control techniques that appear to qualify as a natural approach by virtue of meeting these criteria, namely, *risk assessment* and *technical performance measurement*.

Risk assessment is a natural function of system management.

Technical performance measurement is a natural way of involving engineering design and test personnel in the system management process by asking them to supply basic facts and data without asking them directly to pass judgment upon their overall significance. Properly developed, an integrated approach to risk assessment and technical performance measurement can be beneficially employed, especially if they are firmly established as complementary activities early in system concept formulation. For convenience, we are going to discuss risk assessment and technical performance measurement separately, and then tie the two together and discuss their combined utility for system management control.

Risk Assessment

By its very nature, risk assessment is an important judgment which is made initially as a consequence of system analysis during system concept formulation. From the standpoint of the system design and development process, risk assessment is first a *technical* evaluation of the engineering capability for achieving desired performance characteristics in a system design in relation to fulfilling its defined mission objectives. This includes an assessment of not only the current technological capabilities, but also the prediction of the probable timely attainment of any additional needed capabilities. Secondly, the assessment of technical risk must be converted to an estimate of system development risk by evaluating the impact of the current and predicted technical performance capabilities upon development costs and schedules.

Assessment of technical risks is always in terms of "degree of probability," and of "confidence" in the probability estimates. Seldom, if ever, are they definable in specific terms as derived directly by analyses, experiments, or tests. As a basis for making management judgments concerning integrated technical performance, development cost, and schedule risks, it is important that the following conditions be kept in mind at all times:

1. A realistic estimation of risk is attainable only when a system

performance requirement has been clearly defined. The performance requirement should be stated in terms of desired operational capability and not in terms of system design details. This does not mean that a firm statement of a performance requirement or operational capability must remain sacrosanct and invariable for the life of a system design and development effort. If the requirement changes for any reason at all, the risk assessment must be reviewed, and if necessary, revised. Very often the process of accomplishing a risk assessment leads to modifying the estimate of risk by either inducing a change in the plan of attack to lessen the probability of design and development risks, or modifying the performance requirements so that the probability of its attainment by following a given system design approach is increased.

2. Risk assessment means taking a *negative* viewpoint in analyzing and estimating the probability of success for given plans of attack for solving system design and development problems. It is concerned with estimating and judging the degree of probability that a specific interplay of technical performance, cost, and schedule requirements *cannot* be achieved if a particular planned course of action were to be followed. An assessment that a given approach involves a "high risk" means that the "best" estimate for achieving success, or specifically for achieving desired technical performance and its impact upon development costs and schedule, is unacceptable in terms of meeting established program requirements (that is, of attaining a specified performance capability by a given date and within a prescribed cost ceiling).

Making credible risk assessment judgments in the process of planning for the development of large-scale, complex systems is a tough methodological problem. Every design parameter is a probability function. Every component, subsystem, and ultimately total system performance output is a probability function resulting from an interaction of probability functions. Systematically identifying the "weakest" links in the interacting probability functions is a meticulous time-consuming process. Successful risk assessment is dependent upon doing so.

It must be stated that the history of the development of complex systems has only demonstrated a universal weakness in system management to accomplish and respond to risk assessment. Of particular importance in accounting for this weakness is the nature of the competitive environment which has been inherently inimical to attaining objective realistic judgments. Increase in complexity of systems expands their costs. Large-scale system design and development organizations with their self-perpetuating bureaucratic baronies are indigenously wasteful of ecomonic resources. Monetary inflation over the period of several years required to design, develop, and deploy a system further contributes to the increase in costs. In order to successfully compete for available funds, original system design and development cost estimates and schedules are consciously made more optimistic at all levels of decision-making authority than could have realistically been justified. Once funded, the chances that a system development effort would be continued by paying the increased costs to obtain delivery of a system have historically been proved very good. "Cost overruns" have seemingly become an accepted way of doing business. The only hope that this uneconomical approach can be changed lies in developing and applying reasonably objective risk assessment and technical performance measurements within the framework of an objective management viewpoint which seeks to establish "facts" rather than "fiction."

Perhaps risk assessment, and its complementary function of technical performance measurement, can become the long-sought bridge between systems management and engineering activities to fulfill finally one of the cherished goals of system engineering; namely, to marry the application of technology with sound business management in systems design and development projects. It is a truism to state that successful business ventures are built upon a capability to assess risks properly and to make investment decisions on the basis of backing developments which appear to offer a reasonable chance of paying off. It seems reasonable to assume that for large-scale complex systems requiring government financing to undertake, it is essential to develop the methodology and the capability to apply it for early identification of technological risks and to arrive at reasonably accurate estimates of the chances of surmounting them before committing the economic and human resources necessary for a long-term costly system design and development effort. In the private sector, it undoubtedly will remain just plain good business practice to do so. Increasingly, however, we are witnessing the phenomenon of the government establishing standards and regulations, and monitoring privately developed "system designs," especially in relation to automotive and household electro-mechanical products in the name of consumer protection with regard to health and safety. Accordingly, the industrial community will need to become increasingly concerned with estimating technical risks in relation to potential monetary liabilities for product failure and its feedback impact upon the cost of doing business and sustaining an acceptable financial return for corporate investors.

Expected Outcomes of Risk Assessment

It has become evident from the history of system design and development programs, beginning about 1960, that the existence of a large array of concurrent and interacting areas of risk at the outset of full-scale design and development can cause a system design and development effort to grow substantially in magnitude; to become inefficient; to exceed its planned design and development cost and schedule; and, often to miss badly in terms of its system performance-effectiveness objectives. Furnishing a basis for preventing such deviations in planned design and development goals is the primary purpose of risk assessment during system concept formulation. Properly accomplished, risk assessment will identify and focus system management attention upon technical aspects of a design and development effort which will involve solving state-of-the-art problems and estimates of the impact of such problems upon design and development cost and schedule considerations.

Risk assessment, to be fully effective, must be coupled with *sensitivity analyses* such as are performed during analyses to derive system performance- and cost-effectiveness models and criteria. Here is the first coupling of risk assessment and technical performance measurement during a system design and development cycle. Although we are deferring a detailed discussion of technical performance measurement until later in this chapter, it is essential at this point to understand that from a system management standpoint it cannot be divorced from risk assessment because, by definition, risk assessment is concerned with evaluating the relative impact of concomitant changes in technical performance, cost, and schedule estimates as a multivariate, three-dimensional matrix. The complexity of considerations which may be involved in risk assessment/technical performance measurement is depicted in the form of an analysis breakdown structure in Figure 10. The elements to be involved in a sensitivity analysis are dependent, of course, upon the number and type of system missions to be accomplished and the complexity and type of equipment and service end items required to produce a given performance capability to carry out the specified missions.

Figure 10 has been drawn to show that risk assessment and technical performance measurement are complementary activities. How can they be employed as the prime vehicle for system engineering control? In answering this question, let us omit a detailed discussion of the specific functions incorporated in the analysis breakdown structure shown in Figure 10 and view only the pattern of the functions to be considered from a system management perspective. What is it that a system manager needs to know to exercise control?

First, all system management is a time-based activity. What a system manager needs to know is dependent upon the point in time at which a given system design and development effort is. In the concept formulation stage, he can best control the selection of alternative and preferred candidate design approaches by assessing the relative technical risks involved. In other words, he will be interested in identifying the areas of technological application which appear to present a high degree of technical uncertainty. It must be assumed also that there will always be unanticipated unknowns which will reveal themselves in any design and development approach which may be chosen. Technical performance measurement is important to the system manager during the concept formulation to the extent that proposed technological approaches have a history of having been used in previous system developments, and consequently there is a record of their successes and failures under environmental use conditions which may be similar to that to be encountered by the proposed new system design and development effort.

As a result of risk assessment during concept formulation, the system manager needs to arrive at as clear a picture as possible of any inherent risk relationships among the costs, schedules, and system performance parameters of the candidate alternative design approaches. He can more accurately do this if he can have the results of sensivity analyses involving parameters such as shown in the analysis breakdown structure of Figure 10. The sensitivity analyses should establish the relationships between the parameters employed for assessing estimated system technical performance and those used for establishing the probable degree of design and development risk for preferred alternative design approaches. The number and kind of interacting analyses to be accomplished must be selected on a judgmental basis. No two systems can ever be handled exactly alike for the purpose of making such analyses. The real worth of the analyses, however, is highly dependent upon the objectivity and creativity with which they are accomplished. If appropriate analyses are selected and refined iteratively, they can constitute the essential cornerstone for attaining realistic identification of areas of technological certainties and uncertainties. The total universe of technological uncertainties can never be completely identified except through repeated test and employment of a system under intended environmental use conditions.

There are skeptics, of course, who will maintain that a system design approach is entirely a product of the system project manager's predilections based upon the character of his technical knowledge, past experience, political and economic motivations, and the biased trust he places in his most intimate friends and advisers. If true, then he will merely interpret the results of sensitivity analysis to his own liking, or to the liking of his trusted advisers, in arriving at risk assessments. It is true that meaningful

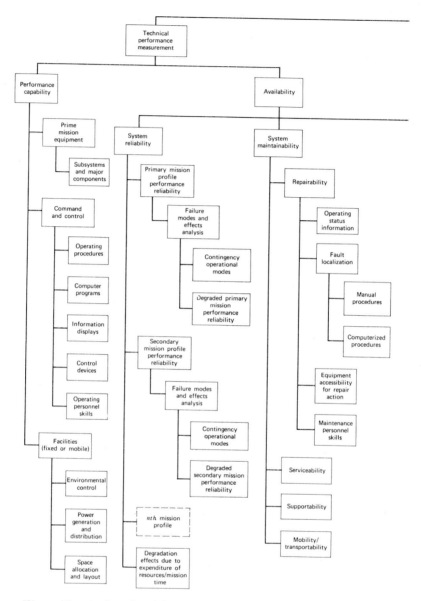

Figure 10. Analysis breakdown structure for risk assessment/ technical performance measurement during design and development of prime mission equipment and associated direct operational support system elements.

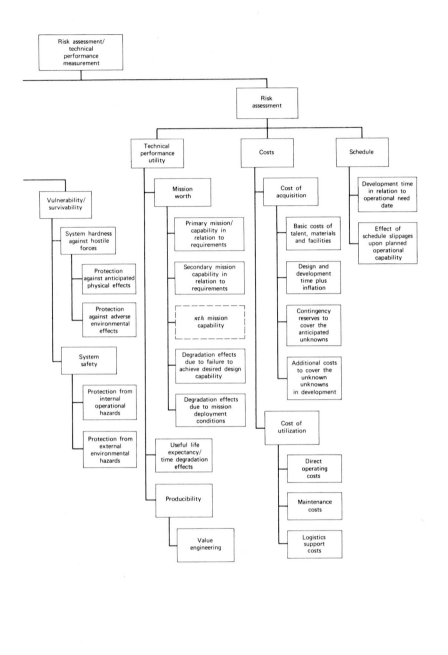

risk assessment must be the product of competent, experienced *judgment* carried out within the mainstream of the system design and project management activities. The best we can hope for is that a system manager would be hard to convince except when he obtains sensitivity analyses, simulations, experiments, or test results which are thoroughly objective and which have been performed under rigorously controlled conditions with the utmost integrity and candor.

Methodology of Risk Assessment

A commonly accepted and tested method of accomplishing risk assessment has not as yet evolved. The approach to be described in this section, and it is just that—an approach only, is based upon logical considerations and a basic methodology which is scientifically oriented. As we have mentioned earlier, risk assessment is first of all a technical evaluation of the likelihood of attaining a desired performance capability by pursuing a given design approach within any specified design and development cost and schedule constraints. To further consolidate and more systematically present some of the general considerations we have been discussing, the information flow diagram shown in Figure 11 has been drawn to illustrate the sequence of functional steps involved in risk analysis, and to incorporate a scale for assesssing technical development risk. It also emphasizes that the process is a closed-loop iteration of analyses and tests. Again, we are further emphasizing that risk assessment and technical performance measurement are complementary functions. That they are is especially borne out as a system design and development proceeds. As engineering development hardware becomes available, emphasis in technical performance measurement shifts from reliance on analyses, simulations, and experiments to demonstrating performance capability under use environment test conditions. Risk assessment becomes less and less a judgmental estimate and more and more a realistic evaluation employing actual statistical data.

In addition to depicting the essential functional information flow for integrating risk assessment and technical performance measurement, Figure 11 has also been constructed to furnish a basis for quantification of risk. The table in the upper right quadrant of the figure sets forth a basis for scaling the relative degrees of technological uncertainty for undertaking a new system design and development effort. In this table the conditions which characterize the highest degree of uncertainty are described in the top row and progress to those which characterize the lowest degree of uncertainty in the bottom row. Note that there is no system design and development effort which can be characterized as a condition where "no risk is involved." Just the very fact that at a minimum, any new system, or even a system modification, involves combining equipment or service end items whose individual functional capabilities may be well understood but when employed in new functional combinations, their resultant integrated performance capability is always in doubt until demonstrated. Estimation of probable functional success of the new combination may be optimistic with a reasonably high degree of statistical confidence, but the exact performance capability cannot be guaranteed until the new system design is actually tried out under its intended use conditions.

Other essential parameters for risk assessment are also shown on the flow diagram in Figure 11. In addition to establishing a confidence level in system performance capability for risk assessment, it is also essential to determine the criticality value of failures in relation to mission functions. To establish such criticality values requires the following steps:

1. The functions which comprise the system mission or missions must be described as discussed in Chapter 5. The functions must be assigned values in terms of their relative importance for fulfilling each mission objective. The breakdown of the system elements, subsystem, and ultimately all equipment and service end items must then be related to each function in terms of the role they play in the successful accomplishment for each function.

2. A *failure mode and effects analysis* is employed to establish the relative significance of the role that each system equipment and service end item plays for the purpose of deriving a criticality value of end item failures in terms of resultant loss in mission

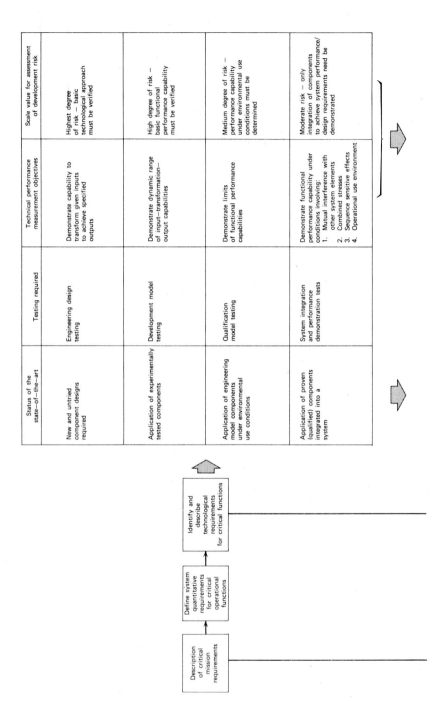

Status of the state-of-the-art	Testing required	Technical performance measurement objectives	Scale value for assessment of development risk
New and untried component designs required	Engineering design testing	Demonstrate capability to transform given inputs to achieve specified outputs	Highest degree of risk — basic technological approach must be verified
Application of experimentally tested components	Development model testing	Demonstrate dynamic range of input–transformation–output capabilities	High degree of risk — basic functional performance capability must be verified
Application of engineering model components under environmental use conditions	Qualification model testing	Demonstrate limits of functional performance capabilities	Medium degree of risk — performance capability under environmental use conditions must be determined
Application of proven (qualified) components integrated into a system	System integration and performance demonstration tests	Demonstrate functional performance capability under conditions involving: 1. Mutual interference with other system elements 2. Combined stresses 3. Sequence sensitive effects 4. Operational use environment	Moderate risk — only integration of components to achieve system performance/design requirements need be demonstrated

Description of critical mission requirements → Define system quantitative requirements for critical operational functions → Identify and describe technological requirements for critical functions

Figure 11. Risk assessment/technical performance measurement information flow diagram.

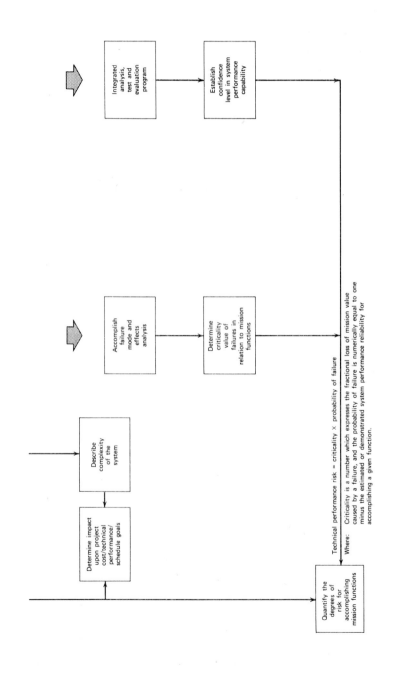

Integrated analysis, test and evaluation program

Establish confidence level in system performance capability

Accomplish failure mode and effects analysis

Determine criticality value of failures in relation to mission functions

Describe complexity of the system

Determine impact upon project cost/technical performance/ schedule goals

Quantify the degree of risk for accomplishing mission functions

Technical performance risk = criticality × probability of failure

Where: Criticality is a number which expresses the fractional loss of mission value caused by a failure, and the probability of failure is numerically equal to one minus the estimated or demonstrated system performance reliability for accomplishing a given function.

performance capability. This determination can be combined with the component end item reliability estimates or performance data to establish a *risk index*. For example, if a particular equipment or service end item is essential for accomplishing a given mission function, it would be assigned a criticality value of 1.00 (where 1.00 equals total failure of mission function). The probability of failure is a numerical value which is 1.00 minus the reliability value (a value, therefore, which will range between 0.00 and 1.00). The risk index is arrived at by multiplying the figure for the criticality value by the figure for the probability of failure. Accordingly:

(a) The *criticality value* is the number which expresses the fractional loss of a mission value caused by a function failure. Establishing such values means that someone must assign a figure of merit for each mission function in relation to achieving specified mission objectives. Such figures of merit will range from 1.00 to 0.00, where 1.00 represents a total mission failure if there is a given function failure.

(b) The *probability of failure* is a figure which is derived from end item reliability analyses and ultimately from test and operational use performance data. It is expressed as the value of 1.00 minus the reliability figure for an end item. The level of equipment and service end item components to which probability of failure values should be assigned is a matter of judgment. Generally, for risk assessment it is sufficient to carry the analysis to the level which identifies the "weakest links" for evaluating a system design approach for performing specified functional sequences. A "weakest link" could be the proposed use of an equipment end item or a subsystem for which the performance capability is marginal in terms of output or useful life, or it could be the proposed use of some part, material, or process with characteristics which make the appropriateness of the proposed application suspect in relation to the use conditions to be encountered. A human function essential to successful mission accomplishment could also be rated as a "weakest link" and for risk assessment purposes could cause a system design approach which depends upon human performance for accomplishing mission critical functions to be viewed with suspicion, especially if the proposed human performance demands appear to exceed human capabilities.

Why are we incorporating this discussion of quantification of risk in a chapter presumably covering "system engineering control"? The answer is quite simple, although conceptually it may take a little thinking about to place in proper perspective. The primary role of system engineering considerations as an aspect of system analysis during system concept formulation is to conceive and evaluate alternative system design approaches for attaining specified mission objectives. If a candidate system design approach should appear to involve one or more "weakest links" as revealed through the kind of risk analysis that we have been describing, then the system engineering control function is exercised in the process of furnishing project management an objective analysis as a basis for arriving at a technical performance risk assessment. When high technical performance risk also involves a high investment in money, time, and human resources for undertaking system design and development, then project management is in a good position to adopt an appropriate strategy for achieving program objectives. Such a strategy can involve a mix of engineering feasibility studies; trade-off studies of alternative design approaches; design and test of experimental hardware ranging from components, through subsystems, to full-scale system prime equipment prototypes; and, engineering development prototypes of operationally configured prime mission equipment, including, if necessary, any associated direct support equipment, facilities, computer programs, trained personnel, procedures, and logistics spares support.

Finally, as a system engineering control function, quantification of risk for risk analysis and assessment such as we have been describing can result in a natural straightforward way of identifying system design and development problems. The results of risk analysis can be presented in matrix form such as is shown in Figure 12.

Matrices, such as are illustrated in Figure 12, can be carried to the level of equipment indenture required to provide complete visibility for risk assessment. Although the analyses may become detailed, the presentation of results in a matrix form permits a system manager to quickly identify and comprehend where the

System level

System segment \ Mission functions		Function A	Function B	Function C
Prime mission equipment	Cr Pf R	1.00 0.35 0.35	0.93 0.25 0.23	0.73 0.10 0.07
Command and control	Cr Pf R	1.00 0.02 0.02	1.00 0.15 0.15	etc.
Facilities support	Cr Pf R	etc.	etc.	etc.

Prime mission equipment level

Subsystems \ Mission functions		Function A	Function B	Function C	
Subsystem A	Cr Pf R				
Subsystem B	Cr Pf R				
Subsystem C	Cr Pf R			Legend:	
Subsystem C	Cr			Cr — Criticality value Pf — Probability of failure R — Risk	

Figure 12.   Illustration of risk analysis matrices.

critical design and development problems reside. He can then concentrate on understanding the technical reasons in explanation of the highest risk indexes. From the system engineering control standpoint, the important principle is that the advantage of risk assessment as a management tool is lost unless the relative importance of technical performance risks can be displayed in a total system context.

Referring again to Figure 11, additional quantification of risk can be made by taking into consideration the levels of technological uncertainty involved in the design and development of various system elements. For example, if a given end item is considered essential for accomplishing a critical mission function (that is, it has a criticality value of 1.00), and because of its design characteristics it is also estimated to have a high probability of failure, then if it is also an item which will require maximum design and development effort (that is, involves the highest level of technical uncertainty), it could logically be assigned a weight of four for multiplication with the risk index. Other end items with perhaps a risk index of the same relative order of magnitude might be rated as falling within one of the other levels of design difficulty or technical uncertainty as described in the descending rows of the table in Figure 11, for which ratings of 3, 2, and 1, respectively, would be appropriate for arriving at a risk assessment in relation to undertaking an engineering design and development effort.

One other factor, namely, complexity of the system, can also be taken into account if it is deemed necessary to do so in order to develop a complete overall system design and development risk assessment. It is possible, of course, that all end item designs could fall well within the characteristics described on the bottom row of the table in Figure 11, but the total number of subsystems or end items which must function effectively together may be such as to justify an estimate that the given system design and development effort presents a large number of technical uncertainties, and, therefore, involves a high degree of technical performance risk.

Collecting Engineering Data for Risk Assessment Use

The system engineering control function via risk assessment methodology cannot be realistically exercised unless the input data are valid. Achieving credibility in risk assessment for complex systems is not an easy task. The necessary judgments must be initiated by knowledgeable end item design specialists at the working level. Their assessments must be filtered and refined by subsystem and system level experts with a broader knowledge and understanding of the physical and functional interactions of all equipment and service end items which are to be employed as an integrated system entity in its intended use environment to attain desired system performance capabilities. The importance of this *total* system approach as a basis for establishing system engineering control of analyses and data compiled for use in deriving risk assessments cannot be stressed too often or too forcibly. It only takes a few moments to state it, but it requires considerable system engineering experience to fully understand it and appreciatively apply it. The nature of the process, inherently involving as it does the development of human understanding of the process, and the intelligent communication of perceived content and the comprehension of judgmental conclusions, collectively constitutes the highest order and, therefore, most difficult scientific and logical process that there is to be mastered. It is never perfectly understood and uniformly applied by all persons who are of necessity engaged in a system design and development effort.

Furthermore, there are so many seemingly incidental and petty political, social, and economic circumstances which can defeat it. People, by their very nature, can never really function in an entirely logical manner in a purely intellectual environment. At best, they can only approach a totally objective viewpoint. The very process of participating in a system engineering or a system management effort causes an individual to identify and become emotionally involved in it. As the process becomes identified in his thinking as a "control" or "forcing" function, it is apt to become immediately stigmatized from his viewpoint as a potential threat to his social and economic security. Intellectual integrity is readily compromised. Judgments are biased in the direction of substantiating a line of endeavor which the participating individual believes offers the most personal security and the least chance of "upsetting his apple cart." True, this atttitude leads to establishing and being satisfied with only short range goals. But, considering that the average span of system design and development is anywhere from about three to seven years, most project management personnel can afford to adopt the short range viewpoint because the expected length of employment on a given design and development project is normally only about two to three years.

It is not the generally accepted style of engineering a large-scale complex system to turn the job over to a relatively small highly integrated team to initiate and carry through its design and development to operational deployment. Rather, a relatively large number of scientific and engineering specialists are employed, and the particular mix of specialties will change as system design and development progresses through the successive stages in its life cycle. Only a very small number of "key" top tier system project management personnel are likely to stay with a given project to guide a system design and development effort from its inception to full maturity. It is because a large number of loosely allied specialists who are normaliy mobile in their work orientation are employed in a system design and development effort that formal system management and engineering control procedures have been invented. Such procedures attempt to substitute a regimented and even a mechanized means of accounting for and tracing design decisions which a relatively close-knit dedicated system design group can achieve through their collective comprehensive personal knowledge resulting from the continuity of design and development effort achieved through day-to-day, person-to-person contacts of the team members. In contrast, response to the imposition of proceduralized approaches lacks the spontaneity and personally felt pride which is generally associated with the self-generated controls that evolve out of a true team effort. There is nearly always a sense of integrity and sense of achievement which is generated by a continuing team effort which cannot be achieved via formalized control procedures. We are, of course, deliberately ignoring the fact that poor project teams can also be responsible for inappropriate or inadequate system designs, and, therefore, on occasion the migration of more competent individuals into such teams can be beneficial.

The migratory individuals who compose large everchanging system design and development project teams generally have the same personal qualifications and motivations as do the non-migratory individuals who compose the small stable project teams. As individuals, they will prefer to exercise their own self-generated judgments and control over their assigned work tasks. When pressed to give a judgmental estimate of the probability of achieving technical performance goals within cost and scheduled constraints, they are naturally going to react—either conservatively or optimistically, each depending upon his own management control viewpoint. If they feel that they are competing for economic support, they will tend to be optimistic about being able to achieve performance goals within the prescribed cost and schedule constraints. If they feel they are to be judged on technical competence, then they are going to estimate the required costs and schedule to be higher than they really believe will be necessary. One may always welcome having a "cushion" available if unanticipated trouble pops up.

In examining the various techniques which system management has attempted to employ to gain visibility and control over a design and development effort, the cornerstone of the most effective approach seems to be the structured interview employing a programmed checklist specially designed to elicit the maximum information and to stimulate personal discovery in the context of given design and development objectives. Emphasizing that we must be concerned with the "multilevel" process of collecting, analyzing, and evaluating engineering data for risk assessment, structured interviews and programmed checklists can best be devised by employing a system analysis viewpoint. Their purpose is primarily to obtain "raw data" concerning technical risks from various engineering design specialists. Such data can then be processed and converted to system engineering risk assessment for presentation to responsible project management personnel. Final judgment and direction of the course of design and development efforts must be their responsibility accompanied by the authority to ensure that it is properly implemented. (As we shall soon see, obtaining visibility as a basis for exercising control of the actual design and development effort is the proper role of technical performance measurement.)

It is necessary that the interviewers be specifically trained and become skilled in dealing with responsible design engineers representing varied technological specialties and design disciplines. Depending on the particular situation, it sometimes helps to establish an effective dialogue between the interviewer and interviewee by asking the latter to fill out a printed questionnaire or checklist as a basis for introducing the interview. More searching, "branching type," programmed questions can be handled effectively by the interviewer in the course of the dialogue, provided he has established good rapport with the interviewee. When an interview is properly conducted, the interviewee 'will gain confidence in the objective viewpoint and attitude of the interviewer and will intuitively recognize the permissive nature of the dialogue. Any implied threat of criticism or punishment if he reveals his true estimate of any design and development problems will most likely cause the interviewee to become evasive and he will then generally try to conceal any judgments he may have made concerning the improbabilities or uncertainties for achieving specified design goals. It must be remembered that in the existing system design and development project environment, our interviewee will likely consider himself to be mobile insofar as employment opportunities are concerned. He most likely will feel that he will be able, if necessary, to avoid the ultimate showdown which will occur when it must be demonstrated that the product of his design effort will perform as required. Consequently, it is essential that the interview-checklist technique be carefully designed so that information can be assembled as "data bits" in such a way that it offers the greatest possibility of verifying the credibility of the answers through the nature of their internal consistency—so that the information obtained is kept relatively free of personal bias or circumvention.

A sample checklist approach suitable for use in interviewing design engineers during concept formulation is shown in Table 3.

TABLE 3. A sample checklist for use in collecting engineering data as a basis for risk assessment

---

Assessment of factors affecting proposed design
and development effort

Project _____   Date _____

System segment _____   Ident. no. _____

Subsystem _____   Ident. no. _____

Configuration item _____   Ident. no. _____

1. Performance/design requirements—The primary performance/design requirements as set forth in Specification No. _____
    are as follows:

    List the performance/design requirements according to their relationship to the present state of the art.
    Requires major advance (≥ 1 order of magnitude):

    Requires nominal advance (≤ 1 order of magniture):

    Within existing technology:

2. Complexity—Based upon previous experience and knowledge of similar systems equipment, the relative complexity of the proposed development effort is:

    ☐ High   ☐ About average   ☐ Below average

3. Technology—In order to meet the specified performance/design requirements, which of the following will be required for engineering development?

    ☐ Simulation programs   ☐ Development engineering model
    ☐ Breadboard/brassboard   ☐ Environmental test chamber
      laboratory model
    ☐ Qualification model   ☐ Production model
    ☐ Prototype model
    ☐ Other special test facility _____

4. Margins—Considering the performance/design requirements and

TABLE 3. (*Continued*)

the available technology, what is the likelihood of demonstrating required performance within the proposed development schedule?

☐ Very good   ☐ Probably can be achieved
☐ Not possible, because _____

_____

(Describe the specific performance parameter(s) and the reason why it is considered that it can not be demonstrated within the proposed development schedule).

5. Tests—Which of the performance/design requirements will require a special test effort to demonstrate because of the inherently critical nature of the design? _____

_____

Is the special test required to demonstrate a capability which is:

☐ Critical to subsystem performance?
☐ Critical to integrated system performance?

Is the requirement for special testing judged to be needed because of:

☐ Uncertainties in the "state-of-the-art?"
☐ No valid reliability data on parts or components?
☐ Uncertainties concerning probable mutual interference problems (electromagnetic, thermal, aerodynamic, etc.)?
☐ Unknown sequence-sensitive effects?
☐ Potentially severe combined stresses?
☐ Untried environmental use conditions?

Can the test(s) be accomplished:

☐ Early   ☐ Midway   ☐ Late

in the development schedule? (If more than one performance/design requirement requires a special test effort, please furnish the required information about each on a separate piece of paper.)

Can the other performance/design requirements be satisfactorily demonstrated by analysis and/or routine verification tests?

☐ Yes   ☐ No

TABLE 3. (*Continued*)

If "no," state exception and describe reason for it.

_____

_____

6. Human resource capabilities—For undertaking the engineering development of the proposed system segment/subsystem/component item, are the appropriately qualified and experienced manpower:

☐ On hand and immediately available?
☐ Available, but will require release from other projects?
☐ Not available and would have to be recruited?

7. Schedule—Considering the estimates of the above factors, and assuming that all test objectives are attained without need for redesign and retest, how many months can it be assumed will be required to demonstrate the specified performance/design requirements?

Number of months at the 80% level confidence in the estimate that all primary performance/design requirements will be demonstrated _____ .

This estimate is at variance with the proposed development schedule by + ____ months or − ____ months.

Name _____

Position title _____ Date _____

Risk Assessment/Technical Performance Measurement During System Design and Development

An integral aspect of system engineering management during system design and development must be a way of monitoring, analyzing, evaluating, and maintaining effective control over all aspects of the design and development process with a view toward ensuring that the resulting equipment and service end items will properly integrate to produce the specified system performance capability. The natural way of exercising system engineering management control during engineering design and development is to devise and conduct a program of technical performance measurement and to review the results of such a program in relation to confirming or revising the established project risk assessment estimates. As design and development proceeds, and technical uncertainties are resolved, the need for risk assessment for system management control is lessened. Even though we desire to emphasize that risk assessment and technical performance measurement are complementary activities and, therefore, cannot really be divorced from a system management viewpoint, for convenience in discussing their relative roles and application, and to tie in best with the discussion up to this point in this chapter, we will finish the discussion pertaining to risk assessment before taking up technical performance measurement.

Risk Assessment During Preliminary Design and Development

During the formal preliminary design task which normally initiates an engineering design and development phase, the previous work accomplished during concept formulation, including such things as advanced system design and definition, the sensitivity analyses, the trade-off studies, the system definition, the end item performance and design requirements, and the program planning—all must be reviewed from the standpoint of whether they truly reflect an integrated system approach. When the system engineering management task is accomplished properly and translated into appropriate direction for the various detailed subsystem and end item design activities, the formal preliminary design review becomes the single most important milestone in a system design and development project. Evaluation of technical risks as identified up to the preliminary design review and of the planned steps to resolve the underlying design and development problems must be a primary emphasis. This review also offers the first realistic opportunity to review plans for technical performance measurement in relation to monitoring progress in resolving technical uncertainties. As a result of preliminary design efforts, and the increased knowledge and insight into the technological requirements and capabilities for translating end item performance requirements into hardware designs, a degree of specificity should exist at the time of preliminary design review which will permit firming up detailed risk analyses and concomitant technical performance measurement plans for monitoring and evaluating the resolution of technical uncertainties and to deal with any unknown unknowns that may be revealed as system design and development progresses.

Technical performance measurement and risk assessment must be made an integral part of the formal design reviews, starting with the preliminary design review and concluding with the critical design review which is held as a basis for releasing detailed engineering drawings and process specifications for fabrication of end items for performance test and evaluation. Such design reviews will cover all prime and support equipment, facilities, computer programs, procedures, personnel training, and logistics support design and development efforts. A technique which is effective in the conduct of design reviews is the use of a design checklist. Such a checklist should be specifically formulated and tailored to systematically cover and interrelate the several equipment and service end item designs which normally are encompassed by a system design and development project. Sometimes this kind of activity is referred to as *interface control,* and may also involve preparation of special interface control drawings to depict the interrelationships of end item physical and functional design characteristics.

The primary purpose of a design review is to demonstrate through presenting analysis and test data that designs in progress are fulfilling specification requirements. When the evidence elicited through the systematic checklist inquiry indicates that design deficiencies exist, then the reason why and the prediction of how and by when they will be overcome become the basis for making a risk assessment. When objectively made, it should result in an adjustment of the integrated technical performance, cost, and schedule estimates, and become a revised project management baseline for the conduct of any subsequent design reviews.

Technical Performance Measurement During Design and Development

Design reviews are conducted against *end item* performance/ design requirements specifications. The proposed design solution for each performance/design specification requirement and its substantiating engineering analyses and test data are systematically reviewed. In connection with such reviews, technical performance measurement should be concerned with evaluating the adequacy of the proposed equipment and service end item designs in relation to satisfying the *system* performance requirement specification in terms of projecting their ultimate functional integration into the intended system configuration. Technical performance measurement, therefore, becomes the technique for carrying out the system engineering management control function in relation to evaluating the design and development effort.

It usually happens at the design reviews that there will be individual end item designs which will be evaluated as failing to meet some discrete performance/design requirements. For a particular end item, the design deficiency may seem to be not too important in terms of significantly degrading its specified input– transformation–output performance capabilities. However, perhaps when viewed in relation to its intended role in conjunction with other end items for producing a required system performance, the design deficiency may seriously endanger that performance. An important function of technical performance measurement is to identify and flag out the importance of such a design deficiency, and to cause a design review action item to be prepared for analysis of its full system impact and recommendations for corrective action. This procedure is especially important for design deficiencies in end items which failure mode and effects analysis have shown to be critical for successful mission accomplishment.

The technical performance measurement function should be covered as an integral aspect of a design review checklist. The criteria for evolving the appropriate checklist items for technical performance measurement are that questions should be structured with a view toward:

1. Evaluation of available engineering analyses and test data for evidence that critical end item performance/design requirements are or are not likely to be achieved within established project cost and schedule constraints. (Critical end item performance/design requirements should normally have been identified prior to a design review in the course of accomplishing failure mode and effects analyses and concomitant technical risk analyses of the proposed system design approach.)

2. Systematically reviewing engineering data for its significance in relation to predicting the probable satisfaction of overall system performance requirements. This should yield a *technical* risk assessment which, in combination with results of the evaluation described in item 1 above, in correlation with other project management functions concerned with maintaining an integrated technical performance, cost, and schedule control, should form the basis for deriving a *project* risk assessment, which in turn is a basis for:

(a) Redirecting the design and development effort (either at the system level or for a specific end item);

(b) Modifying an engineering analysis or test program;

(c) Improving a project management function; or,

(d) Revising project technical performance, cost and schedule estimates.

One particularly important payoff for the results of technical performance measurement in projecting probable system performance capabilities will be the information that it can provide as a basis for refining cost and schedule estimates. Just as in the beginning of a design and development effort the estimates of ultimate system performance capabilities are probabilistic in nature, so are the estimates of system design and development costs and schedules. Here the so-called "iceberg" phenomenon occurs. There is a larger number of technological unknowns which occupy the area below the visible surface at the start of the project than the number of known technological solutions appearing above the surface. By definition, it is the purpose of an engineering design and development effort to resolve the technological uncertainties, both anticipated and unanticipated (that is, "known unknowns" and the "unknown unknowns").

Since system design and development cost and schedule estimates have little credibility unless they reflect what is really needed to support a design and development effort, such estimates must necessarily be subject to revision as a consequence of planned technical performance measurements and risk assessment. Initial system design and development cost and schedule estimates should always realistically indicate that a considerable gap exists between the most "optimistic" and "pessimistic" estimates. As design and development proceed, and verified knowledge about system technical performance capabilities becomes more precise, the probability that technological uncertainties still exist will be progressively lessened. Consequently, the cost and schedule estimates to cover remaining design and development tasks can become more precise. Revision of project cost and schedule estimates on the basis of technical performance measurement and risk assessment at the time of design reviews should be just as important as are changes in design approaches.

Methodology of Technical Performance Measurement

Theoretically, at any point in time during system design and development, it should be possible on the ·basis of all available engineering data to project the probable performance capability of a system as the end result of the on-going design and development process. However, it does not appear realistic to attempt to compile the kind of comprehensive data and to keep it updated on as frequent a basis, perhaps daily, as would be necessary to provide a capability for an instantaneous appraisal of a system state. Conditions are normally such that the periodic scheduled design reviews are time-phased to best fit the scope and depth of a given design and development effort so that they are naturally the most appropriate times for attempting a formal assessment of the "system state."

Figure 13 depicts a functional definition of what constitutes a "system state" for purposes of technical performance measurement and risk assessment for application to in-process design and development efforts. Interest in describing the system state on

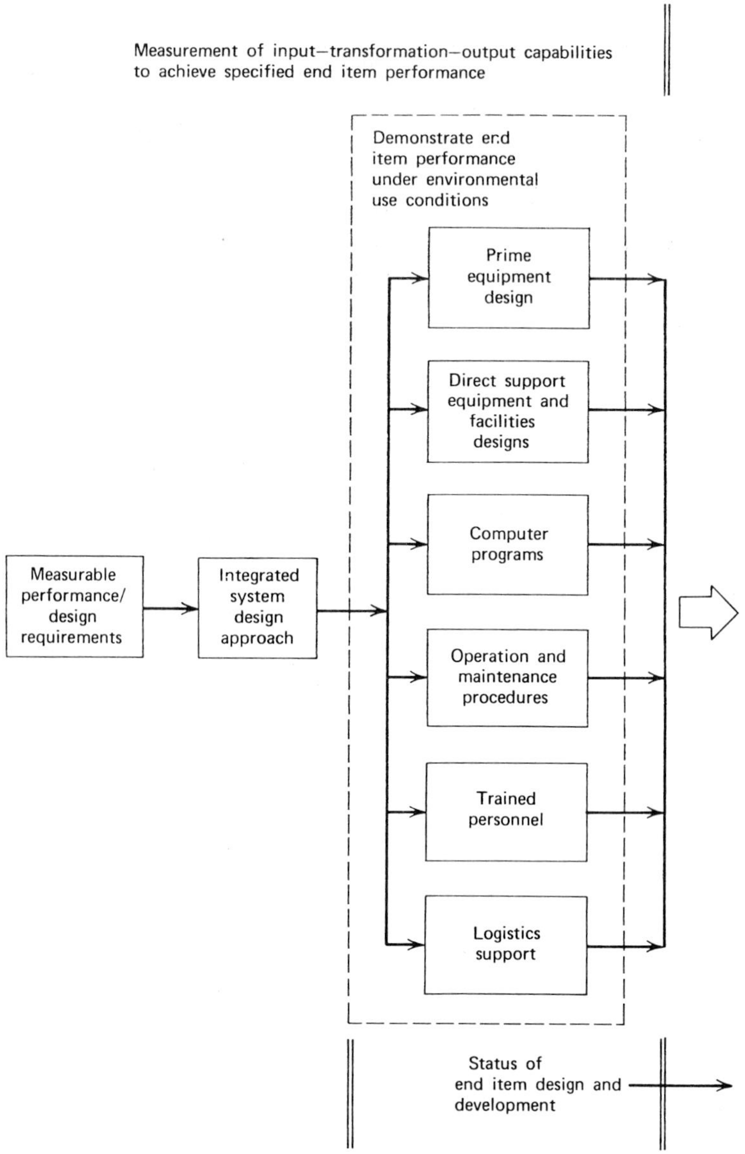

Measurement of input—transformation—output capabilities
to achieve specified end item performance

Demonstrate end
item performance
under environmental
use conditions

Prime
equipment
design

Direct support
equipment and
facilities
designs

Computer
programs

Measurable
performance/
design
requirements

Integrated
system
design
approach

Operation and
maintenance
procedures

Trained
personnel

Logistics
support

Status of
end item design and
development

Figure 13. Definition of "system state" for use in technical performance measurement.

the basis of technical performance measurement centers on verifying predicted events and in revealing (as rapidly as possible) unanticipated events. This is essential to progressively increase the confidence level in the validity of the estimate of

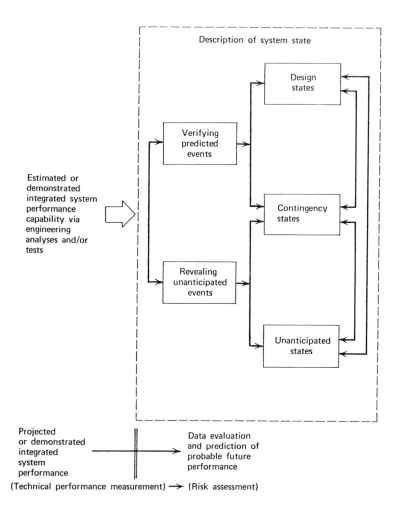

probable system performance effectiveness as derived from any given analysis of a current design state. In Chapter 6, we discussed an integrated analysis, test, and evaluation approach for system design and development which is essential as a basis for providing the engineering data needed for determining technical performance measurements. For project control purposes, system engineering, and in turn system management, need only be concerned with evaluating data bearing upon those parameters which have been established as involving high technical risk and, consequently, implying potential cost and schedule risks. Of course, at any time during system design and development, the results of analyses and testing may reveal that something which had been previously classified as a "known" technological capability can migrate into the "unknown" category. Continuous analysis is essential during system design and development to detect such shifts and to modify technical performance measurement objectives accordingly. As has been discussed in connection with risk analyses, selection of the relatively few significant or critical performance parameters for monitoring is the key for successful system engineering management control of engineering design and development.

It must be stressed that technical performance measurement just for the sake of making measurements, or confining technical performance measurements to those parameters for which it is relatively easy to compile quantitative data such as weight, space, power consumption, and fuel consumption, can be essentially a sterile effort from the standpoint of the objectives of system engineering management control. Compilation of such data is important for configuration management control of the design requirements allocated to end items, but technical performance measurement to be useful to a system manager must be sensitive for revealing unwanted deviations in the planned design state which potentially will jeopardize the attainment of specified system performance requirements. Such deviations should be capable of early detection and identification as a design problem to be resolved as soon as possible. The effectiveness of technical performance measurement, whether it be adopted as a formal or an informal procedure, therefore, must be judged by the capability it provides for early detection of design problems likely to have an adverse effect upon overall system performance, cost, and schedule objectives.

The category "contingency states," shown on Figure 13, covers those performance anomalies in a system which, if they occur, can result in a degraded performance capability and may require either a design modification or a "work around" procedure to restore some semblance of a specified normal mode of operation. When a system is committed to an operational mission, the capability to institute contingency procedures is very valuable when certain performance anomalies occur. It may mean the difference between a mission success, or partial success, and a failure. Technical performance measurement, therefore, should be as equally concerned with demonstrating contingency performance capabilities for key system parameters as with verifying full normal performance effectiveness.

Even in what might be considered as the ultimate verification of a system design state by repeated operational application .of a system to accomplish its specified mission, the probability that unanticipated events will not occur is never reduced to zero. There is, accordingly, always a possibility that there are residual unanticipated states in a system design that can reveal themselves at some future time. There is a saying that, "If anything can go wrong, it will."—Murphy's Law.

The Pitfalls of Trend Analysis

One device commonly advocated in connection with technical performance measurement is the use of "planned parameter profiles." The process of selecting these profiles starts with the compilation of a master parameter measurements list which contains all technical performance measurement requirements for each of the system design parameters.

For each of the system design parameters, a planned parameter profile graph is prepared with a heavy solid line representing the projected time-phased demonstration of a performance requirement. This line drawn in relation to the calendar date baseline presumably locates expected performance measurement goals of the design and development effort for the selected design parameters. These graphs are intended for use in relation to comparing "measured progress in performance" with "planned performance" at the designated milestone dates. The "measured performances" are to be entered on the graph, and usually are connected by a dash line for comparison with the "projected performance measurement" shown by the solid line. The usually stated purpose of this tracking device is to provide system management with the visibility by which they may forecast system progress and detect possible trouble areas. These are brave and noble purposes, if they can be achieved.

Most plots of "achieved" versus "planned" values, usually accomplished on a monthly basis, suffer from the following defects:

1. The determination of what "achieved value" to enter on the graph for a given month will be some engineer's estimate of how well a given design approach will satisfy specified performance requirements. This estimate is subject to the following kinds of biases:

   (a) Responsible design engineers will always "hedge their bets." Estimated "achieved performance values" on a monthly basis permits them to report "poorer than expected" one month so that they can report "better than expected" the following month. The overall effect is generally to make their progress report look good to their system project manager. They don't lose anything by reporting an "underachievement" for a given month, especially if it is early in design and development effort, because there is still plenty of time remaining until their hardware or software end item must actually be tested. Reporting an "overachievement" likewise makes them look good because it leaves the impression that they are "on the ball" and are making their design goals ahead of schedule. Monthly consistency in under- or overestimating actual against planned achievement is hardly ever attained.

   (b) The system performance parameters chosen for "tracking" may be those for which it is easiest to derive an estimate on the basis of objective design analyses and component test data such as weight, hydraulic power demand, electrical power demand, and propulsive thrust. It is essential, of course, that there is accountability in the allocation of such parameters in order to maintain control over the overall end item performance/design characteristics. However, such system performance parameters as range, reaction time, signal strength, availability, maneuverability, and vulnerability/survivability are those which are characteristics of a "system" design resulting from the integration of equipment and service end item designs which cannot be tested until the system as a whole is actually committed to a test. It is unrealistic to try to obtain overall "achieved performance values" for these system parameters on a monthly basis by using "engineering estimates."

2. Trend analysis is a management gimmick which by its very nature invites "gamesmanship." Until it is necessary to conduct an actual performance test, a design engineer is reasonably safe from criticism or censure in reporting values that he thinks management would like to hear from their particular political or economic viewpoint at the time an estimate is made. If the actual test to be conducted some months later indicates a better performance than predicted, it only makes the responsible design engineer look good. If there is a failure to meet performance specification, it probably can be rationalized away as being due to "unanticipated" problems arising out of system 'integration."

In the final analysis, the best in-progress technical performance measurement will be a judgmental estimate by competent system engineers who are familiar with the system, and with the various equipment and service end item performance/design requirements. System engineers should be capable of reviewing, and evaluating engineering design and test data, and projecting it in terms of probable system performance effectiveness in its intended use environment. The more these estimates are based upon test data from experimental or prototype models of components, subsystem, or complete systems, the more accurate will be the prediction of the performance effectiveness of the ultimate operational system which is evolved as a consequence of a well-planned engineering design and development effort.

The Environment for System Engineering Control

Instead of introducing the subject of system engineering management and control in the customary manner of starting out with an overview, we are going to conclude the discussion in that manner. We will use the graphic illustration shown in Figure 14 as a basis for the discussion.

To facilitate the discussion of the interrelationship of system engineering with other major activities involved in a total system design and development effort, we are going to restrict ourselves in relation to a system life cycle to the concept formulation and engineering design and development phases where system engineering presumably plays its most important role.

The illustration in Figure 14 depicts the relative levels of effort for the various activities with which system engineering must interface during system concept formulation and engineering design and development. For the purposes of further simplifying the discussion, we are including as an aspect of concept formulation the step of formalizing a system design and definition as a basis for undertaking an engineering design and development effort. Many system project management charts and discussions will show this as a separate phase. This is appropriate from a project management viewpoint, but for engineering purposes it does not need to be treated as a separate phase, since it is only a task prerequisite for transitioning from concept formulation to actual design and development of system equipment and service end items. Also, the wavy line shown separating concept formulation from engineering design and development is intended to

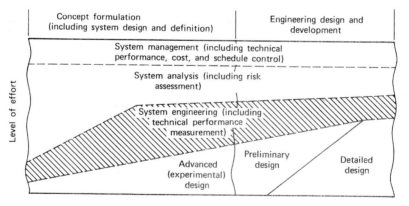

Figure 14. Interrelationship and relative emphasis of system design and development project activities.

indicate that there is no clear-cut borderline between the two phases. Since the total system design and development effort is an iterative process, not all concept formulation will necessarily be complete prior to undertaking engineering design and development, and the design and test of engineering models may cause some of the effort to revert to further concept formulation. But, assuming that the unusual really happens and that everything proceeds as planned, and in summary of our discussion on system engineering management and its role in the overall system management, the chart has been drawn with the following considerations in mind:

1. Overall system management must be concerned with control of the expenditure of money, time, and human resources to achieve a desired system performance effectiveness. Consequently, system managers are interested in having visibility of the progress toward achieving desired technical performance, cost, and schedule objectives. In this effort they are in turn dependent upon the quality of data furnished by system analysis. The principal system management control device is the work breakdown structure and monitoring progress in design and development efforts in relation to the planned scheduling and pricing of the inclusive work packages.

2. System analysis is primarily concerned with selecting the basic system design approaches for satisfying the established mission requirements. System analysts must be concerned with relating the proposed system design approaches with the overall design and development objectives and with ensuring that they are cost effective as well as performance effective. Risk assessment is employed to identify state-of-the-art problems requiring special consideration in establishing technical performance, cost and schedule goals and for determining technical performance measurement requirements.

3. System engineering is concerned with specifying and monitoring the design and development of equipment and service end item designs which will properly interface with each other to produce a *total* system performance capability for attaining specified mission objectives. Technical performance measurement based upon an integrated analysis, test, and evaluation program

designed to progressively predict and demonstrate system performance-effectiveness goals can be an effective system engineering control device employed for monitoring key indicators of the progress in the system design and development effort.

Why hasn't this overall structuring of system management and system engineering in relation to end item design and development activities ever really been successfully implemented? First of all, the need for an extensive system management effort in relation to producing mainly government-sponsored large-scale complex systems with arbitrarily imposed cost and schedule constraints has only existed since about 1960. Secondly, the record of excessive costs, schedule slippages, and failures of systems to achieve performance expectations during the decade 1960–1970 adequately attests to the fact that system management and system engineering schemes largely invented and presumably implemented during that time period have not fulfilled their promised goals. Why not?

The hard lesson which should have been learned is that a ready-made system management structure incorporating a rigorous system engineering procedure cannot be superimposed upon already established ways of doing business with the expectation of immediately and succesfully supplanting them. People change their attitudes, viewpoints, and habits only gradually. Attempts at forcible change through introducing documentation requirements are resisted and only make the process of transition from old to new ways of thinking and working more difficult to accomplish. Furthermore, overdocumentation in an attempt to introduce system engineering procedures as a "forcing function" in the basic system and equipment design effort has been resisted as an attempt to regiment the thinking of people who normally consider themselves to be essentially creative, independent, and well-qualified specialists in their chosen field. On their part, the management "experts" in self-defense have become "cultists" and tend to view the universe of system design and development procedures and techniques as their special province. They soon adopt a strict constructionist perspective and become more concerned with the refinement of a formalized procedural approach and associated documentation techniques than with increasing their utility for achieving a cost- and performance-effective end product. In view of these historical facts and the resulting insight gained, what are some guidelines for achieving a successful system engineering approach for a design and development effort?

1. System engineering considerations must be applied as an integral aspect of system analysis early in system conceptual formulation. It becomes increasingly difficult to successfully introduce design changes as a result of system engineering considerations the farther a system progresses into ever more detailed design and development efforts. Once a given configuration has been tested and accepted, design changes are most difficult to justify in terms of improving system cost- and performance-effectiveness.

2. A system project management framework built upon a foundation of objectivity and technical integrity, and which is favorable to the application of the system aproach, must exist in order to provide an environment favorable to the use of system engineering methods and techniques. Conversely, the more that a system management perspective is dominated by political, social, and economic considerations rather than by technical ones, the less a truly objective system engineering approach is likely to succeed.

3. A system engineering (or a system management) procedure cannot be successfully promulgated for immediate application. A period of trial introduction and gradual implementation in convenient steps is essential in order to achieve the requisite understanding, confidence, and acceptance by those personnel responsible for carrying out the details of a system design and development effort. Most people on a large-scale system design and development project do not normally achieve a system-level "big-picture" viewpoint concerning the ultimate importance of the work that they are doing. Deliberately communicating and instilling a system viewpoint is a prerequisite for expecting project personnel and supporting functional area specialists to effectively work together toward a common set of system design and development goals. Otherwise, "gamesmanship" can be expected to be the normal way of doing business. Furthermore, ensuring that personnel assigned to system project offices restrict their

efforts to solving system problems in support of the attainment of detailed equipment and service end item designs by appropriately specialized design groups will go a long way toward establishing effective working relationships between the two different levels of endeavor.

4. A system design and development approach based upon the design and test of a progressive series of equipment prototypes is more amenable to the application of the "system approach" than is reliance on voluminous, and sometimes seemingly interminable, paper design analysis and trade-off studies. For a given system design and development effort, the application of the appropriate "system approach" should result in the proper mix of analyses, trade-off studies, and component prototype hardware design and test. Such an approach incorporates the natural problem-solving approach essential to the application of the scientific method and must be encouraged. Actual experience in large-scale complex system design and development projects has furnished ample evidence that it is the most cost-effective approach in the long run, and results in the attainment of the desired system performance effectiveness more rapidly than by following the route of attempting to design and test all elements of a new system "concurrently" in order to produce it in quantities for expedited operational development. When a decision to produce has been based largely on engineering analyses unsupported by adequate test data, the consequences have been that design deficiencies have been revealed in operational test or use which degrade the system performance effectiveness and require expensive retrofit design modification and retest programs in order to attain the desired or necessary system performance effectiveness.

5. Finally, acceptance of all of the above considerations will still fail to achieve the desired objective of attaining effective system management and system engineering practices unless project managers and their advisory staffs are properly qualified and oriented to apply a system approach for decision making during system concept formulation or engineering design and development. We are now expounding the reprise of the basic theme of this book, namely, increasing the effectiveness of system engineering management can be achieved only by improving

the capabilities of responsible project and direct support specialist personnel for identifying and dealing effectively with system problems. It can never be achieved solely by improving the mechanics of documenting management schemes, that is, by continually refining the "folklore approach" to system management. With this thought, we are ready to discuss the career development needs of system project managers.

10

Developing
System Managers

Effective System Design Starts with Top Management

It is a common phenomenon when a system design and engineering effort is in trouble for project management to recruit specialists in the hope that they can stem the tide of disaster. More often than not, specialists are the one kind of personnel that are least needed. It is a reasonable assumption that when this happens, it is a clear indication that the system project management is hardware-oriented rather than system-oriented. In such a case, however, it would do little good to bring in highly qualified system design and engineering personnel. Unless top management in a system project office has the basic system-oriented philosophy and understands the methodological approach requisite to translate it into action, personnel working for them who are system-oriented and skilled in systems design and engineering techniques will be working in a climate of frustration. Without effective system-oriented management leadership, understanding, endorsement, and active support, their efforts can only be an exercise in futility. Top management must not appear to be a talent inversion in the eyes of the subordinate members of a system project office.

If we ever hope to achieve the scientific approach to system design and engineering that we have been describing, it is essential that a carefully prepared educational program be organized with a view toward developing the requisite knowledge and skills for successfully undertaking system-level engineering and management of design, development, and testing activities. Only by such a concerted educational effort to develop compatible system engineering and system management skills can we hope eventually to organize like-minded and properly qualified personnel at all levels of responsibility in system project management organizations.

Doctrine of Relevance

As we have attempted to show, system-oriented design and engineering is primarily attained by acquiring a system-level perspective. Major effort in an educational program, therefore, must be placed on getting system level management and engineering personnel to *learn, feel, understand,* and *appreciate* how system design is spawned, nurtured, and produced.

Actually, there is no known basis upon which simple statements can be prepared and set forth in a checklist form to be used as a self-instructional program which will automatically cause management and engineering personnel to gain the understanding and insight necessary for recognizing and defining those things which constitute system problems, and then for thinking about and evaluating alternative ways of solving them in order to evolve a synthesized sytem design solution that will satisfy the requirements which originated the problem. How do you adequately describe and communicate to others the way to become a creative thinker? Those who have a record of sucessful scientific discoveries, inventions, product development, management improvement, and the like, understand the problem-solving and decision-making process and can communicate effectively with each other about its characteristics. They have a common perspective. Without a mutually understood common perspective, it is not possible to develop a sense for what is relevant for solving system problems which normally involve many complex, artificially contrived and interacting parameters. When one hears the comment, "You make system problems too complicated," it simply means that the person making the comment does not understand the nature of systems or the approach to designing and engineering them.

In attempting to encourage specialists to become system engineers by motivating them to master and be devoted to the system-level decision making, it is necessary to do more than exhort them to adopt a system viewpoint. It is essential to provide appropriate case material and guided experience. It cannot be expected that a technical or design specialist will shed his hardware-orientation in a hurry. It takes time and patience. It is a case of providing the opportunity—from the top system project management down to each system design and engineering staff member—for learning to discriminate between those matters which are appropriate to be acted upon at the system level and those which should be referred to the hardware level for decision making in regard to appropriate detailed design implementation. System-level designers and engineers, of course, must be continually concerned with verifying that detailed engineering properly translates system requirements into equipment designs, but the decisions on how to accomplish such translation are best made at the detailed design level.

Table 4 compares the difference in system *versus* equipment design viewpoints for the important system performance- and cost-effectiveness parameters. It brings out as clearly and succinctly as possible the essential differences between the design viewpoints for the two approaches. Both are valid, of course, at the respective levels of interest. Implementing requirements for each of the factors at the equipment design level without first establishing those at the system design level is characteristic of hardware-oriented system design approach. Such an approach has been characteristic of many system design and development projects. As a consequence, if the equipment items could be made to work at some specific time and place, the results were considered acceptable. The task of integrating the resultant collection of end items so that they work together as a functional entity to achieve a system performance capability has placed an overwhelming burden upon the test and field engineering personnel. Numerous "make-it-work" design modifications usually must be made. This approach usually adds both costs and time to system development. Only through the development of appropriate system design and management skills can it be hoped that system definition and design will effectively lead the hardware design

TABLE 4. Relative Emphases in System *versus* Equipment Engineering Viewpoints

Factor	System Design Approach	Equipment Design Approach
Capability	Operational utility	Performance output
Reliability	Man-machine dependability	Failure-free use cycle
Maintainability	Ease of maintenance to increase availability	Packaging for servicing and repairability
Cost	Optimum performance effectiveness within available resources	Application of most economical materials and processes to attain performance requirements
Schedule	Time to define and evaluate interactions of all components	Time to design, fabricate, and test deliverable end items

effort. To be effective, system managers must be skilled in restricting attention to those things which are relevant for determining system designs in relation to specifying performance and design requirements to hardware design and development managers and designers.

Throughout our discussion we have deliberately avoided the use of material illustrative of the detailed application of the system approach. Such illustrations would pertain to the *mechanics* of system design and engineering and we did not desire to distract the reader's attention from the *dynamics* of the process. The *dynamics* of system design and engineering cannot be illustrated but only described. Understanding and insight into the appropriateness or the suitability of a given methodological approach or the use of specific techniques can be gained only through experience. Each application is unique unto itself in a given context. It is not truly generalizable to all system design and engineering development projects. However, the use of the scientific frame of reference as the universal approach to describe objectively and analyze system problems and to select potential solutions for design, development, and testing is generalizable. Experience with applying the scientific approach to system design and engineering results in increased understanding of its applicability. The writer has been involved in implementing the system design and engineering process for a number of projects. Each one has contributed further refinements to his understanding of the *dynamics* of the process. That is the way human learning at the level of insight occurs. It is a process of personal discovery that no amount of the mastery of the use of the *mechanics* of system design and engineering can bring about without the elements of curiosity, imagination, and insight being present in the individual's behavorial makeup.

Basic Qualifications of a System Engineer

To become a successful project manager it is essential that one first become a successful system engineer. As we have discussed elsewhere in this book, system engineers must evolve out of one of the numerous scientific and technical specialties which are employed in deriving basic system designs. When a specialist demonstrates an affinity to the system point of view, he is receptive toward acquiring the following additional qualifications:

- A capability for conceiving and planning the applications of optimum combinations of equipment, men, procedures, and supporting techniques to achieve some stated goal or purpose.
- A working knowledge of many disciplines such as mathematics, electronics, physics, mechanics, physiology, psychology, computer programming, economics, and so on; plus a mastery of the methodology and techniques for formulating and solving system-level problems.
- Detailed knowledge on how to prepare and manage:
    Technical design and development, and test programs;
    Uniform specification and data management programs;
    Engineering procedures for translating system requirements into end item designs;
    Plans for procurement and manufacturing of equipment end items, construction of facilities, and production of support techniques and procedures;
    Design, development, and test of computer programs;
    Plans for selection and training of specialist personnel to operate, control, and maintain the system hardware and software;
    Project management schedules and progress reports; and,
    Cost estimates, budgets, and controls of expenditures.
- A facility for directing and coordinating the efforts of a multi-specialty system design and engineering project management team.
- A keen sense for sorting and sifting facts in order to select those that are relevant to a given problem and can, therefore, legitimately be employed for effective system design and engineering decision making.
- A sensitivity to the importance of the established political–economic–technical–human factors climate which generates the customer's (whether actual or prospective) value standards by which he will be judging the acceptability of the system product.

Working at the level of a system project office, a system engineer's output will be ideas and the directions required to implement them. Neither will have any force until they are placed on paper. The rewards of seeing hardware or software taking shape and performing successfully as a consequence of the implementation of his ideas and directions will be indirect. From the standpoint of satisfying a creative urge, it is important that system engineering does not become stultified because it is constrained and overburdened with formalized paperwork procedures. Uniformity of procedures is important but their overelaboration to the nth detail is deadly and unnecessary. It discourages effective system engineering and encourages the proliferation of "papermongers," who become more concerned with format than with content, with style than with substance, with procedures than with process, and, consequently, with the *mechanics* of system engineering than with its *dynamics*.

Basic Qualifications of a System Project Manager

In addition to demonstrated success as a system engineer, to become a successful system project manager requires leadership qualifications which are usually found in successful executives everywhere. He should:

- Be confident of his own capability to understand and direct a *total* system development effort.
- Seek out and acquire staff personnel who possess more knowledge in some areas than he does.
- Refuse to allow a rigid approach or too early a commitment to a given design to occur.
- Be comfortable with relatively unstructured situations in which the project personnel with various special talents form and reform into different task-oriented subgroups to deal with the various types of problems which are bound to arise.
- Need very little in terms of status symbols to identify to others his position of authority or of the crutches of formalized procedures to enable him to function effectively.
- Readily identify an appropriate role for himself when partici-

pating with various subgroups, leading some, and being a member of others. He has, therefore, both leadership and membership skills which he applies as appropriate to a given situation.

- Enter easily into group discussions to surface problems and to contribute to problem solutions on an objective basis in a manner that does not leave the impression that some members of the group have "lost" while others have "won" in the decision-making process. He can, therefore, make the final decisions when necessary and win their acceptance without smothering individual initiative which he may need very badly in future problem-solving situations.
- Show confidence in his associates, be sincerely interested in their professional capabilities and rely on them to do their jobs. In the context of system design and development, he helps them grow into more competent generalists and discourages any tendencies to perpetuate specialist elitism.
- Not accept poor or sloppy performance, but know how to convey his dissatisfaction to groups and individuals in a manner that is supportive, constructive, and acceptable, and, therefore, motivates change.
- Respect and encourage the maintenance of technical integrity both for himself and on the part of his staff. He will not tolerate manipulation of himself by other people, nor will he attempt to manipulate them. He accepts disagreement and differences of viewpoint among his staff personnel as assets in pursuing solutions to complex system problems. He is able to assist them to properly formulate the differences and work through them in order to arrive at "best fit" solutions in relation to the system design objectives.

Training for System Engineering and Management

System engineering and management training, of necessity, must be planned on a long-range basis. Short-term attempts to "bootstrap" the caliber of system engineering management cannot succeed in developing the depth of understanding, perspective, motivation, and maturity of experience which are required for providing the desired leadership qualities needed for guiding the development of today's complex systems. The effective amalgamation of the multispecialty system project organization into an integrated interdisciplinary effort in practice continues to be evanescent. This is undoubtedly at least partly due to the lack of a "system approach" to system management. The primary objective of a system management training program should be to implement such a systems approach along the lines described below.

Effective systems management is an integration of talent, leadership style, and organizational climate. In any system design and development effort there are several echelons of organized endeavor which must be employed. For the purpose of this discussion, we are limiting our concern to selecting and training of personnel to handle system project office tasks. This activity, in contrast to hardware design and engineering activities, is, generally speaking, poorly understood. Consequently, tailored training for assumption of system management tasks has been understandably neglected. As a basis for setting forth a "systems perspective" for the purpose of discussing the specifics of a systems management training program, Table 5 contrasts "favorable" *versus* "unfavorable" characteristics of various factors which influence the quality of a *system* design. This table has been divided into: I. Some basic personnel qualifications; II. Management philosophy; and, the integration of these to provide, III. Working environment. With the criteria set forth in Table 5 in mind, there are at least four major things which must be properly integrated in order to achieve a "systems approach" to system management training, namely,

Criteria for selection into training,
The training approach,
Performance evaluation, and
Utilization policy.

Criteria for selection and training. System design and development projects require the services both of scientific, engineering, and management specialists and of system engineering and management generalists. The latter must of necessity be developed

TABLE 5. Factors Influencing Quality of System Design

I. Some Basic Personnel Qualifications

Element	Favorable Characteristics	Unfavorable Characteristics
Intellectual ability	Ingenuity in synthesizing new approaches to solve system problems	Unimaginative application of standard approaches to handle limited system problems
Attitude	Accepts role of independent thinker	Primary concern with acquiring cloak of group acceptance
Language facility	Ability to communicate new ideas	Expresses ideas in a specialist group-oriented jargon
Decision making	Evaluates relative value of problem solutions in terms of probable effect upon system performance	Relates proposed problem solutions to a few fixed value standards regardless of their system impact
Motivation	Devotion to the scientific approach for deriving optimum design solutions	Main concern with increasing level of management responsibility
Temperament	Willingness to make decisions on the basis of incomplete supporting data	Requires conclusive data before committing self to a line of action

II. Management Philosophy

Element	Favorable Characteristics	Unfavorable Characteristics
Line of authority	Well-defined hierarchy of decision-making authority	Indeterminate, vacillating, or delayed decision-making authority
Project organization	Task force approach to project management with directive authority	Loose alliance of highly Balkanized functional groups with no central control

TABLE 5. (*Continued*)

II. Management Philosophy (*Continued*)

Element	Favorable Characteristics	Unfavorable Characteristics
Technical direction	Emphasis upon proper decisions by the right people at the right time	Preoccupation with achieving specialized application of esoteric technology
Product evaluation	Emphasis upon application of functionally integrated cost-effective technology in synthesizing designs	Emphasis upon promotion of specialized cultist approaches as cure-alls for system design problems

III. Working Environment

Element	Favorable Characteristics	Unfavorable Characteristics
Resources	Close-knit design teams with well-balanced, comprehensive technological and management experience	Overloading of system design staffs with prime mission equipment specialists
Communication	Mutually understood and accepted ways of defining, analyzing, and evaluating system-level problems	Employment of functional area specialists to handle system-level problems
Rewards	Recognition for contributing to increased effectiveness of system designs	Making a monetary profit without particular regard to performance effectiveness of system products
Requirements	Well-defined system requirements and constraints by higher management	Generalized, vague, ambiguous, or undefined system requirements
Time phasing	Sufficient time to permit experimentation and iteration of analyses in order to optimize system design	Fixed schedule for production and delivery at the sacrifice of adequate analyses and testing
Reports	**Simple, natural, easy to use formats for** reporting only significant data	Elaborate, artificial, incomprehensible reporting of minutiae

from among the ranks of the former. There are a number of devices which can be employed for selecting prospective system engineers and managers. For the purpose of this discussion, we are not going to attempt to differentiate between a system engineer or a system manager. For all practical purposes their duties and tasks are essentially the same, with the former oriented more toward "technical" management, and the latter toward "program" management. However, this is really more of a difference in emphasis in job duties rather than a difference in prerequisite fundamental knowledges and skills. Both jobs require the same basic orientation toward an understanding of the "system approach." The main difference in the two job efforts is between that of a "technical" or "engineering" interest in designing and producing a system end product, and that of "managing" or "controlling" the overall engineering and production effort within the framework of technical performance, cost, and schedule goals. A common training base is needed for both jobs. The selection criteria can be virtually identical and are generally indicated in Part 1 of Table 5. Identifying basically qualified and experienced scientific, engineering, and management personnel who are potential system engineers or managers should involve the use of a composite of techniques such as:

- *Individual Interview*—Already successful system project managers can be requested to interview and evaluate individuals whose work record and expressed job interests indicate a leaning toward assuming system level duties and tasks.
- *Group Judgment*—A number of individual judgments can be compiled to derive a consensus in regard to a given individual's potential as a prospective system engineer or manager.
- *Tests*—For professional personnel with job histories such as normally will be considered for selection into system engineering and management training, use of psychological aptitude tests would be essentially nonproductive. A simple questionnaire designed to reveal attitudes, interests and motivations in relation to undertaking system-level job responsibilities will suffice to supplement or to lead into individual interviews such as are described above.

- *Training Courses and Job Trial*—Assuming that all indications are favorable, or even that a given individual insists on volunteering for system-level training, the best selection device is a formally organized training course and subsequent job trial. Even for those who will subsequently decide that their individual skills and interests are best utilized in pursuing their basic technical speciality, their system-level training will be beneficial inasmuch as they will better appreciate how their specialty can be effectively applied in relation to achieving overall system design and development objectives.

The training approach. A certificate of successful graduation from a single training course will not suffice to produce a "system engineer" or a "system manager." Becoming a successful system engineer or manager requires a continuing learning process. Beginning the process in relation to personnel who possess only basic technical and/or management knowledges and skills, sequentially the steps in the system-level educational process would be essentially as follows:

- *Familiarization or Orientation Training*—This would be a brief one to two week course of one hour per day with visual training aids and reading material designed solely to present the "systems approach" with a view toward broadening an individual's perspective and to identify and define to him the nature of "system problems" versus specialist hardware and/or technical program problems and their systems applications.
- *Intensive Formal Training Course*—For those who elect to pursue the subject beyond the orientation course, a formal course, approximately eighteen weeks in length, meeting for a two hour session each week (or equivalent) would be conducted to provide systematic coverage of system engineering and system management methodology and techniques. Teaching would be conducted by lecture and discussion based upon textbook reading and study of case materials. Practical exercises employing simulated system planning, design, and development problems would be conducted. Major emphasis of this course should be on developing a substantive understanding

of the systems approach, rather than with extensive explanation of programmatic details.

- *Supervised Job Experience*—Following successful completion of formal training, or even concurrently with the formal training course, depending upon available openings in project office staffs, arrangements would be made with the responsible project managers to schedule periodic seminar-type discussions among course graduates, either individually or in groups, to take up such system engineering and management problems as they may wish to volunteer for discussion. It would be desirable that attendance at some designated minimum number of such seminars, say over approximately a two-year period, would lead to the granting of an advanced certificate in system engineering and management.

- *Special Briefings*—From time to time government agencies will adopt new policies and issue new directives or instructions affecting the conduct and reporting of contractual system engineering and management procedures. Some of these changes will be of such a scope and will have such a major impact upon established system engineering and management procedures that a special familiarization briefing should be conducted for all project management personel who may manage any new project which will be covered contractually by the new system engineering and management requirements.

- *Technical Specialty Courses*—To support system-level engineering and management, a number of engineering and management technical specialists are usually required, e.g., data management, cost and schedule control management, configuration management, design reviews, functional analysis, interface control, test and evaluation, reliability, maintainability, value engineering, safety engineering, and so on. Specialist courses should be conducted to prepare project office personnel properly to carry out the various techniques in a manner to best satisfy contractual requirements. The requirements and techniques for such activities are subject to relatively rapid change. In some cases, therefore, it may be necessary to conduct special indoctrination in relation to each proposed new system design and development effort in order to properly prepare given technical specialists to implement the new system management procedures.

- *Cooperation with University Training Programs*—One important area which has not been adequately covered in the past is the feedback of verified system engineering and management practices from industry to the universities and colleges which will be furnishing the graduates who will become tomorrow's system engineers and managers. How best to effect a closer liaison between industry and university teaching staffs needs to be explored. Perhaps familiarizing them with, and even inviting them to participate in a company's in-house training effort could be the constructive channel of communication. An information bulletin written specifically to be informative concerning representative system engineering and management problems, and the methodology and techniques employed in handling them, could possibly prove to be beneficial and could also serve a useful company public relations purpose.

Performance evaluation. For a training program to have a payoff, both for the individual and for a company, it should be coupled with some scheme for performance evaluation. The classical, and verified approach would be to: (a) Establish standard position descriptions for various system engineering and management jobs; (b) Employ position-oriented performance standards as a basis for evaluating the output of job incumbents; and, (c) Maintain a periodic rating and evaluation of individuals who constitute the pool of qualified system engineering and management personnel. To be beneficial for both properly utilizing the available skills and talents for maximum individual job satisfaction and for keeping a company in the best position for conducting system design and development efforts, ratings and evaluations of system engineering and management personnel should be specifically tailored to identify both their proper job placement and their growth potential. The factors to be considered should cover a sufficient number of aspects of an individual's performance in order to properly reveal, and above all else, to convince him of his true strengths and weaknesses in relation to the types of system engineering and management functions and positions which are available, or likely to be available in the future. The factors to be evaluated must be pertinent to the actual work of system engineers and managers and should be directly observable. For the level and type of personnel who will normally be employed in a system engineering and management effort, self-ratings are feasible as well as peer ratings. A few, but significantly diagnostic, factors can suffice for rating and evaluating the performance of system engineering and management personnel. The ultimate objective should be for the individual to accept an objective assessment and to adjust his level of aspiration to best capitalize upon his strengths and to avoid being placed in job situations where his weaknesses will most probably lead to something less than successful job performance. No performance evaluation scheme can ever be perfect, but if properly adapted to given job situations, it can be a valuable management tool. Without such a scheme, there is no basis for judging the payoff of a training program such as we have been discussing, because there is no way of objectively following up personnel upon completion of a unit of training in order to assess whether their productivity over a period of time is judged to be relatively successful or unsuccessful as a feedback for possible improvements in the training program.

Utilization policy.   System engineering and management by the inherent nature of the activity requires a product-type organizational structuring of the jobs to be accomplished. To properly qualify personnel to undertake system- or project-level tasks, the training program must be organized to prepare specialists drawn from various subject matter disciplines to take a team approach to problem solving in relation to system design and development goals within any established program constraints. Personnel who are successful at accomplishing this are going to be identified as being competent in a subject not as yet fully recognized as an academic subject matter or discipline. They essentially must forego any interest in maintaining an identification and seeking an increasing status recognition within a functional scientific, engineering, or management specialty grouping. A company, therefore, undertakes a special obligation to properly utilize qualified system engineering and management personnel once it has invested in their future through a company sponsored training program. In fact, once it has put together a winning project management team, from an economics standpoint it behooves a company to keep it together for employment on new projects. The best predictor of success is demonstrated success. Furthermore, the uncontrolled migration of personnel between functional specialty groups and project management staffs should be discouraged. It is best that an individual elect either to remain as a technical specialist or to become a system-level generalist as a full-time occupation.

Many specialists attempt to make the transition to become qualified as generalists for system engineering and system management activities. Most of them fall back to pursuing their basic scientific or technical field as being more appealing or interesting to them. This process of "natural selection" is very desirable. It allows those with the real talent for system design and engineering to be identified and to develop. It furnishes those who elect to continue in their specialty field the opportunity to become acquainted with the system approach and, therefore, to be more receptive to system-level direction of their efforts with a view to producing end items which fulfill specified requirements for proper system application.

Organizations which have a history of producing successful system designs keep dedicated winning system engineering and system project management teams together to work on successive system projects. They appreciate the fact that the success of their products is dependent upon the close-knit team work of such a group. The familiar old adage is: "Nothing succeeds like success."

Epilogue

We have tried to take a "systems approach" in dealing with the substantive issues which determine the quality of system designs which emerge from given design and development efforts. Pragmatically, in systems acquisition there has been a strong inclination to build the hardware first, and to give thorough attention to the operational requirements only when the whole system begins to collapse. In an attempt to counter this tendency, programmatic techniques have been devised with a view toward providing better management visibility and traceability of interlocking system design and development decision making. The difficulty of this approach has been that it has identified "system engineering" with elaborate documentation schemes to the detriment of focusing attention upon developing system managers who are effective system engineers because of their system-oriented viewpoint, talent, and leadership capability for amalgamating the efforts of many specialists in order to rationally derive a cohesive system design.

Effective leadership of a system design and development effort exists only when operational expectations and values are demonstrated by system performance and not by achieving conformance with programmatic procedures such as may be incorporated in various "management systems." Documentation schemes can only provide the semblance of system engineering, not the substance. A competent system manager will know all about the status of his particular design and development effort. He will be well on the way to solving any problems as they occur before any management system will have had time to document them in order to provide him with the "visibility" that they exist. A data compilation "system" cannot be substituted for knowledgeable and alert people when it is desired to achieve an effective surveillance over a design and development effort. A management information processing system may conceivably span geographical separations and provide an aid to the distant people who are required to evaluate information, to exercise judgment, and to make authoritative decisions. However, personal contacts are always faster than any techniques of documentation.

Effective system design and development efforts require a personal engineering style which can be stifled by the adoption of elaborate systems of documentation. The objectives that documentation schemes are intended to attain must be ingrained in the thinking and motivation of people who manage a system design and development effort. When they are, the need for the documentation as a basis for decision making effectively disappears. However, the documentation of decisions after they are made should be the basis for directing and controlling the efforts of those who will be implementing them via the various hardware, software, and support service end item design and development efforts which, when properly integrated, will constitute the system.

Effective system engineering, therefore, is not the result of any specific procedural approach and documentation scheme, but rather it is the product of:

1. The particular talent and capability of the people who are placed in management control of a system design and development effort.
2. The style of engineering leadership as it impacts upon welding together a multispecialty effort to produce a coherent and unified system end product.
3. The degree of authority and responsibility that a given system manager has for making and controlling system and end item design and development decisions.

Bibliography

Chestnut, H., *Systems Engineering Methods*, Wiley, New York, 1967.

Chestnut, H., *Systems Engineering Tools*, Wiley, New York, 1965.

Eckman, D. P., ed., *Systems: Research and Design*, Wiley, New York, 1961.

Ellis, D. O., and F. J. Ludwig, *Systems Philosophy*, Prentice-Hall, Englewood Cliffs, New Jersey, 1962.

Gosling, W., *The Design of Engineering Systems*, Wiley, New York, 1962.

Hall, A. D., *A Methodology for Systems Engineering*, D. Van Nostrand, Princeton, New Jersey, 1962.

Machol, R. E., ed., *System Engineering Handbook*, McGraw-Hill, New York, 1965.

Sandler, G. H., *System Reliability Engineering*, Prentice-Hall, Englewood Cliffs, New Jersey, 1963.

Shinners, S. M., *Techniques of System Engineering*, McGraw-Hill, New York, 1967.

Index